Three into One
has been published
as a Limited Edition
of which this is

Number 147

The bicentennial commemorative panel, with different elements sponsored by different firms.

THREE INTO ONE

The Three Counties Agricultural Society 1797-1997

BY

JOHN LEWIS

BARON
MCMCVI

PUBLISHED BY BARON BIRCH FOR QUOTES LIMITED
AND PRODUCED BY KEY COMPOSITION,
SOUTH MIDLANDS LITHOPLATES, CHENEY & SONS,
HILLMAN PRINTERS (FROME) AND WBC BOOK MANUFACTURERS

© Three Counties Agricultural Society Limited 1996

All rights reserved. No part of this publication may be reproduced, stored in a retrieval system, or transmitted, in any form or by any means, electronic, mechanical, photocopying, recording or otherwise, without the prior permission of Quotes Limited.

Any copy of this book issued by the Publisher as clothbound or as a paperback is sold subject to the condition that it shall not by way of trade or otherwise, be lent, re-sold, hired out or otherwise circulated without the Publisher's prior consent, in any form of binding or cover other than that in which it is published, and without a similar condition including this condition being imposed on a subsequent purchaser.

ISBN 0 86023 556 4

Contents

Acknowledgements and Sources	6
Foreword by Richard Law frics faav, Chairman of Council	7
Farming Renaissance	8
Livestock and Orchards — The Hereford Agricultural Society	15
Ploughing and Pears — The Worcestershire Agricultural Society	32
Merger in Chamber — Worcester joins Hereford	43
A Respectable Assemblage — The Gloucestershire Agicultural Society	52
Showtime! — The Three Counties Agricultural Society	70
Home, Sweet Home — The Society at Malvern	83
Behind the Scenes	99
Endgame	113
Appendices	120
Index	128
Subscribers	131

ACKNOWLEDGEMENTS

The Minute Books of the Hereford Agricultural Society and the Three Counties Agricultural Society are virtually complete. For the rest I have relied on contemporary accounts written at the time.

I am indebted to Lyn Downes, Tony Halls and the staff of the Three Counties Society for their assistance in searching out old records, to David Burlingham, Gerald Sivell and Dr Celia Miller for pointing me in the right direction at critical moments, to the staff in the Public Records departments and Research Libraries of the three counties, and finally, the reminiscences of 'the Ancients', gleaned in the markets of Hereford, Worcester and Gloucester.

Whilst the author has made every effort to acknowledge material sources, he apologises in advance for any that have been inadvertently missed.

Book cover design by Helen Lewis.

SOURCES

The Hereford Times, series of articles by T. W. Garrold, *The Worcester Journal, The News Chronicle, The Illustrated London News, The Gloucester Citizen and Journal, The Cheltenham Chronicle and Gloucestershire Graphics, The Cheltenham Examiner, The Farmer and Stock Breeder, A History of Worcestershire Agriculture* by R. C. Gaut, *General View of the Agriculture of the County of Gloucester* by Thomas Rudge, *General View of the Agriculture of the County of Hereford* by John Duncomb, *General View of the Agriculture of the County of Worcester* by W. Pitt, *Leominster Guide* of 1803, *Patriotism with Profit, British Agricultural Societies of the 18th & 19th centuries* by Kenneth Hudson, *Social History of England* by G. M. Tevelyan, *Diary of a Country Parson 1758-1802* by James Woodforde, *Foreman of the Fields* by John Lewis, *Yeomanry Cavalry of Gloucester and Monmouth* by W. H. Wyndham-Quinn, RASE, Journal 1901, *The Book of the Farm* by Henry Stephens, minute books of the Three Counties Society, Hereford Society, Hereford and Worcester Society and Gloucester Society.

DEDICATION

To Ivor and Sue whose courage and example are an inspiration to so many.

FOREWORD
by Richard Law FRCIS FAAV, Chairman of the Council

Some 200 hundred years ago in Herefordshire the founders of the Society were concerned with how they could help farmers to produce food of the highest quality at the lowest prices — has anything changed? In those times attendance at shows was an essential source of advice for farmers and the sharing of information. The Society, through its shows, encouraged improvements in agricultural production illustrated by demonstrations of new implements and machinery. Competition by showing of livestock was encouraged together with improved breeding and this led to the establishment of pedigree herds.

The history of the Society has been traced by the enthusiasm and research of John Lewis; fresh from chronicling *A Century of the Cotswold Vale* he has now embraced two centuries of the Society to provide a benchmark in the history and activities of the countryside. The fortunes of the Society have followed those of farming in the rural counties of Hereford, Worcester and Gloucester.

From times of plenty to scarcity those working and living in the countryside visited the show as it moved through the three counties. During the depression of the 1920s the emphasis of the show changed and non-agricultural interest increased attendance. Entertainment was provided in the form of military displays, show jumping and other exhibitions which attracted people from the towns.

The Society looks forward and continues to foster change through learning within agriculture and providing a place for horticulture, crafts and leisure pursuits, furthering the appeal to everyone.

ABOVE: Life on the farm in the mid-nineteenth century: women played a large part in field operations. BELOW: Mid-nineteenth century farming and implements.

Farming Renaissance

The years 1797 and 1798 were some of the most critical in our island history. Napoleon was invincible throughout Europe and was turning his eyes towards England. The French *Directory* had named him Commander in Chief of the Army of England and orders were placed in French shipyards for invasion barges.

At home, partly because of the brutal conditions under which ships' crews were expected to serve and partly in sympathy with the French Revolution, the Channel Fleet mutinied at Spithead and the trouble spread to the Northern Fleet and the Nore. Panic spread throughout the country and there was a run on the banks by farmers and others for cash. This became so widespread that an order was issued authorising the Bank of England to refuse cash payments until Parliament gave further orders.

The threat of invasion continued until Nelson won a resounding victory over the French fleet at the Battle of the Nile and the danger receded.

In the same year, 1798, Hereford held its first show in Broad Street, Hereford. We would have little difficulty in visualising the scene, the minutes and contemporary accounts being similar to the minutes and reports of today's livestock committee.

The conditions in the countryside at that time were somewhat different and bear examination. We can then understand why the fortunes of our societies waxed and waned throughout the nineteenth century and the context in which they developed.

At the turn of the nineteenth century, more than half of the arable land of England and Wales was farmed on the open field, common farm system. Under such an arrangement no individual could do anything unless everybody else agreed. The improvement of livestock was a wasted labour when the half-starved cattle and sheep of the village were running together on the same common with the village bull.

The village would be self-sufficient in everything and such tracks as there were would end at its furthest bounds. The land between villages and towns would be forest and scrubby waste, inhabited by robbers and outlaws. It was always advisable to travel in company.

The effect of the Napoleonic wars and, indirectly, of industrial development was to extinguish the open field system of land cultivation. This resulted in the passing away of multitudes of small landowners and commoners, who were tempted to sell their rights for temporary affluence, quickly dissipated at the nearest alehouse. During this time, 1790 to 1820, it is reckoned that three and a quarter million acres were enclosed and upwards of forty thousand small farms thrown into larger holdings. Contrary to what might be expected, the western counties were quicker to enclose their commons than the eastern counties. Even so, in the early part of Queen Victoria's reign there were still large areas of Worcester and Gloucestershire unenclosed. The parish of Elmstone Hardwicke in Gloucestershire was the last — in 1916.

By 1815 the war was over, the profits and prices of war had disappeared but wartime taxes ands costs persisted. Distress was widespread, rents were substantially reduced or remitted altogether. House of Commons select committees sat almost continuously from 1820 to 1836 inquiring into the causes of agricultural misfortune. The last of these commissions made no report and the impression is that by 1836 the depression was subsiding.

The conditions that prevailed at the time must account for the fact that the records are sketchy. Agricultural societies formed and reformed a number of times. It would also account for Gloucestershire starting much later than the other two.

Large landowners and the nobility were the prime movers in starting these societies. With approximately ninety per cent of farms tenanted it was common for landlords to encourage competition between their tenants. This is especially shown in the records of the Gloucester Society. On some estates throughout the country, landlords gave prizes in the form of bulls. By encouraging the tenants in this way there would be mutual benefit to landlord and tenant.

During the war agricultural wages remained stationery, while prices of necessities rose rapidly. Weekly earnings were supplemented from the rates, the parish guaranteeing 'a living wage'. The payment was increased in proportion to the number of children. Ratepayers burdened by the increasing load of poor relief spent less and less on wages. In this situation Societies offered a prize varying between two and three pounds for the family who brought up the greatest number of children without having to resort to poor relief. With wages between eight and ten shillings, (forty to fifty pence) this prize must have seemed like untold wealth.

Although a Scotsman, Andrew Meikle, built a thresher in 1786, flails were still the universal means of threshing corn. By 1850 steam power became available, which encouraged manufacturers to develop threshing machines. Improved ploughs, reapers, haymakers and other machinery encouraged progress. Even so by 1850 high farming was still the exception; more than half the occupiers had made little advance on the practices of the eighteenth century.

The failure of the potato crop 1845-46 caused a severe famine and led to the repeal of the Corn Laws. When in 1846 protection was abandoned for free trade an agricultural panic resulted. This is shown in the records of the societies, much time being spent in debate on the situation.

Because of protection, competition for farms had been reckless. In 1850 rents had risen a hundred per cent since 1770, while the average yield of wheat per acre had only risen from twenty-three to twenty-six and a half bushels per acre and at £2 a quarter, wheat was the same price as it has been eighty years previously.

Ruin was widespread, land was thrown on the hands of landlords and efforts were made to convert arable into pasture.

Fortunately the situation was short-lived; from 1853 onwards there was a rapid recovery. Gold discoveries in Australia and California raised prices, trade and manufacture boomed, the Crimean War closed the Baltic to Russian grain and farming entered a period of prosperity which lasted to 1874.

During the period 1863-74 improvements in livestock breeding were great and continuous. Stock breeding and 'pedigree' became a mania among men of wealth and the agricultural showyard came into its own. Not only did Shorthorns, Herefords and

Devons attain a high standard, but other well-known breeds were brought to perfection. In sheep the improved Lincolns, Oxford Downs, Hampshire Downs and Shropshires were virtually creations of this period.

Prosperity encouraged the investment of capital at unsafe rates of interest, leading to reckless bidding for land, which raised rents beyond their safe limits.

In America a slump and fall in wages had driven thousands to settle in the mid-west and develop the natural resources of virgin soil. In this country between the years 1875 and 1879 a series of disasters struck. Wet summers ruined the harvest and potato crops, there were outbreaks of rinderpest, foot and mouth, pleuro pneumonia and foot rot in sheep. To cap it all in 1878 the Glasgow, Caledonia, South Wales and West of England Banks all failed.

Steam carriage and low freight charges enabled the world to compete on home markets and between 1875 and 1900 the national wheat acreage fell from three and a half million to one and three quarter million acres. Not only grain but imports of meat and dairy products rose dramatically. The population seemed to buy their food anywhere except at home.

The availability of cheap supplies of food from abroad heralded a decline in farming fortunes whch, apart from a brief fillip during the 1914-18 war, lasted until 1939. During this time, the influence of the landed gentry and aristocracy was eclipsed by the barons of industry.

A period of rural poverty set in, the farm labour force slipped away to the town in increasing numbers, where urban employment gave them better wages and working conditions. Politicians could afford to ignore the situation in the countryside, because no unemployment problem resulted.

Complain as they might, landlords' and farmers' political influence was waning. After a long period of peace since the Battle of Waterloo, war and its accompanying food shortages were a forgotten memory.

Politicians of all parties felt they coud afford to ignore agriculture. They theorised that, if they neglected the land, other industries would make up the shortfall in national income and all would be well. They forgot that farming has a far wider interlinking with other industries.

A worrying result, being increasingly felt up to the present day, is how sadly ill-informed large sections of the population were about the source and vulnerability of their food supplies. Coupled to this had been the decline of a class of person, best described as the independently minded yeoman character, who had served the country so well over many centuries.

As always in adversity the farmer's resourcefulness pulled him through. The development of the internal combustion engine and labour-saving machinery lowered costs of production. The great strides made in all classes of stock breeding from 1850 onwards, were appreciated in the ability of good stock to thrive. Not for nothing was England regarded as the stock farm of the world. The opening of agricultural colleges contributed to the understanding of the science of farming, thus encouraging more economic production.

The agricultural showground provided a valuable stage for the development and demonstration of agricultural machinery. In addition it was essential for competitive development and as a shop window for livestock.

LEFT: Sir Anthony Lechmere of The Rhydd, Worcester supported the original Hereford and Worcester Societies. His direct descendant, Sir Berwick Lechmere, was President of the Three Counties in 1982. RIGHT: Thomas Duckham Esq, (centre left) Secretary of the Hereford Agricultural Society 1874-1881 and prominent Hereford farmers in the 1880s. BELOW: Viscount Cobham, President in 1930 and grandfather of the present Viscount Cobham. OPPOSITE BELOW: His son, Viscount Cobham, was President in 1967.

The progression of what became the Three Counties Agricultural Society continued for well over half a century, despite the disruptive effect of two terrible wars. The Society was able to weather the depression of the 1920s because the emphasis of its Show was subtly changing. Non-agricultural trade stands increased, and military displays, horse-jumping and other attractions in the main arena brought the urban dweller, who came to the show for entertainment. The introduction of the motor car and the assurance of a good day out helped to ensure its popularity throughout the depression years.

The years following the second world war were arguably the best that the agricultural industry has ever seen.

LEFT: The Cotterell family of 'Garnons', Hereford have been supporters of the Society for many generations: Sir John Cotterell, grandfather of the present Sir John. RIGHT: Sir John Cotterell, father of the present Sir John, was President of the Society in 1950.

LIVESTOCK AND ORCHARDS —
The Hereford Agricultural Society

There may be some confusion as to which is the year in which the Hereford Society was founded, 1797 or 1798, the first show being in 1798. However, Rev John Duncumb in his *General Review of the Agriculture of the County of Hereford*, states that — 'an Agricultural Society in the County of Hereford was established in 1797, and it comprises most of the principal proprietors and many of the principal occupiers of land, throughout the county. The number of members at present [1805] exceeds one hundred and twenty'.

As Duncumb was the first Secretary of the Society, we must accept this authority.

Records of the first show are minimal but we know it was held in Broad Street, Hereford in the spring of 1798. This was a cattle show; there was a further meeting on 15 October 1798 when the prizes offered were: six guineas for a new apple, five guineas for ploughing the greatest number of acres with oxen and two guineas for rearing the biggest family without the aid of parish relief.

The Earl of Oxford was the first President and among the early supporters were the Duke of Norfolk, and names that are still well-known, such as Biddulph, Cornewell, Cotterell, Scudamore, Symonds and Stallard.

Prizes for livestock given at the summer meeting of the Society in 1799 consisted of three silver goblets, value five guineas each, for the best bull, not over twenty months, best bull not over three years and seven months and a prize for the best boar pig.

At the autumn meeting of the same year no premiums for livestock were offered, but prizes were awarded for ploughing, turnip crops and a new apple. Other prizes were given for bringing up large families and for long service.

Early in the nineteenth century, when corn had rise to 154 shillings a quarter, because of the Napoleonic wars, the Society induced farmers to send corn to Hereford market to be sold to the poor in small quantities. A premium of a shilling a bushel was also given for the first two hundred bushels of new potatoes sold in Hereford market not later than August.

At the June meeting of 1800 there was, as the result of a private bet, the celebrated contest between Mr Meek's Leicester bull and Mr Purslow's Hereford bull. The stake was a hundred guineas, the judge was Mr Prater, a Somerset breeder, who gave the award to the Hereford.

OPPOSITE ABOVE: The members' Annual Dinner was an important feature of the early agricultural shows. In the absence of cameras this early *Punch* cartoon gives a good impression of the atmosphere. The subject of the address and the fact that the author has heard the same message on at least three occasions during his lifetime should give some comfort to anyone suffering from 'agricultural depression'. BELOW: Enclosure map of the Parish of Elmstone Hardwicke in Gloucestershire.

Evidently the purposes of Agricultural Socities did not find universal favour. Jonathan Williams, writing in the *Leominster Guide* of 1803 stated: 'The objects of this [Hereford] and other societies are certainly of great national importance, but whether the means which have been adopted are best calculated to produce the ends proposed may admit of serious doubt. Every improvement in the science of agriculture should be directed to the reduction of the price of necessities and not to enhance them.
'Individuals and not the public have derived benefit from them. If the interests of the community be consulted no-one will deny that the agriculturist who produces, on a given proportion of land, the greatest quantity of livestock, is more deserving of encouragement than he, who in fattening an animal consumes the meat that would suffice for five.'

He then goes on to make some scathing comments about cross breeding: 'as it is acknowledged by the most scientific that the admixture of foreign breeds with the genuine stock of the cattle and sheep of the district, would prove injurious rather than beneficial'.

A pointer to the Societies' encouragement of the production of new varieties of apple is given in the following passage: 'England's cider apple orchards appear to have been planted early in the reign of Henry VIII. Herefordshire is indebted for all the fine old cider fruits to the industry of the planters of the early part of the last century and the end of the preceding one. By the spirited exertions of Lord Scudamore of Holme Lacy and other gentlemen of the county, it became in a manner of speaking, one entire orchard. All the old cider fruits are now either lost or so far on the decline as to be deemed irrecoverable'.

In another comment: 'A genuine Hereford Ox, the largest ever seen in the Kingdom, far exceeding in size the Durham Ox has been reared by Mr Taylor of Yatton — the present rage for fine woolled breeds of sheep is to be deprecated, neither the pure bred Herefordshire stock of cattle or sheep are capable of further improvement'.

In 1804 four shows were held, also the usual sales, the extra one being at Leominster, where occasional visits were afterwards paid. Prizes were offered for a great variety of objects, including manure, drainage, the comparative value of an acre of hoed turnips and an acre of cabbage, the planting and care of orchards, for the greatest number of hives of bees raised by cottagers and for a quick set hedge.

The Secretary of the Society was one of forty winners out of 350 competitors for prizes offered by the Board of Agriculture, of which the then Lord Carrington was President. The prize was awarded for an essay on 'The best mode of converting grass land into tillage, without exhausting the soil, and of returning the same to grass, after a certain period, in an improved state or at least without injury'.

The Michaelmas meeting of 1805 was by far the largest and most important held hitherto. The entries were a record and furnished convincing proof of the determination of Hereford farmers to keep their beloved white-faced cattle well before the public. A Hereford Ox that year took the first prize of twenty guineas and an additional prize of 10 guineas for the best beast over 120 stones at Smithfield Show. It was bred by Mr E. Walwyn of Marcle, an original member of the Society.

The subscription to the Society was one guinea per person and one suspects that some members, while enjoying the benefits, were somewhat lax in paying up and this accounted for the varying fortunes of the Society. The following table gives some indication of the Society's fortunes during its early years.

Year	Subscriptions	Cr/Dr balance
1799	£99 16s 0d	Cr £43 5s 6d
1800	£63 0s 0d	Cr £35 16s 6d
1801	£96 12s 0d	Cr £29 12s 0d
1802	£59 16s 0d	Dr £3 6s 6d
1803	£75 12s 0d	Dr £38 15s 6d
1804	£109 18s 0d	Dr £42 3s 6d
1805	£118 13s 0d	Dr £53 15s 0d
1806	£164 15s 0d	Dr £1 7s 6d
1807	£108 3s 0d	Dr £29 9s 0d

In 1810 the first meeting of the year was held at the Greyhound Hotel under the Presidency of Lt-Col Foley. The best yearling bull came from Mr Yarworth of Troy, Monmouth, best two-year-old from Mr Kedward of Westhide and the best aged bull from Mr Preece of Lugwardine. The number of cattle exhibited was larger than usual and several yearling bulls sold for £50.

The first of two summer meetings was held on 7 June at the Green Dragon, Hereford when there were classes for yearling heifers, cart stallions, aged ram, yearling ram and best boar pig.

The second summer meeting was held at the King's Arms, Leominster a week later on 15 June. There were classes for the best milch cow, three-year-old bull, yearling heifer, three-year heifer, boar and ram.

2 July 1810 was the date of the annual wool sale. The wool was brought into the city to be sold and it was found to be too wet to hold the sale out of doors, as was usual, so the Society obtained the use of the Shire Hall. Prime Ryeland wool fetched 42s a stone. On 31 August 1810 the annual ram sale was held under the auspices of the Society.

At the first meeting of 1811 there was a discussion on the best method of catching rats. This was again referred to on 6 June and there was correspondence with the Bath and West Show about a Mr Broad's method.

This same year awards were given for long service, first prize-winner being John Jones aged 78, for 65 years' service at Castleton Farm, Clifford.

The October meeting was attended by most of the nobility and gentry of the county, including Lord Oxford, Lord Hereford, Sir George Cornewell, Sir H. Hoskyns, Sir G. Cotterrell, Col Scudamore MP, Mr Anthony Lechmere of Worcester, Mr Westcar of Bucks and Mr Tench of Shropshire.

At a meeting on 3 February 1812 there was correspondence with the Lords of the Admiralty about Mr Broad's method of catching rats.

It was also proposed to publish a list of defaulting subscribers, in the hope that the publicity would shame them into paying up.

There was considerable discussion on the uniformity of the bushel measure. Apparently it was passed into law in 1197 in the reign of Richard I, that the country should adopt 'The Winchester Bushel', as a uniform measure. Such was local custom that it had still not been universally adopted.

A special meeting was called on 14 May 1813 'For the purpose of expressing approbation of the proposed manner of making public Mr Broad's "receipt" for destroying rats'.

Mr Knight reported that he had tried out the method and was convinced that rats could be destroyed by it under almost any circumstances.

A most unsuccessful meeting was held in October 1813 when only two prizes were awarded, 'demonstrating the decadence into which the Society had fallen'.

However, the new President, Hon A. Foley MP, put new life into the Society and the meeting on 7 February 1814 was reported as one of the most satisfactory ever held.

A vote of thanks was proposed to an unnamed schoolmaster of Llangollen who, by cutting grooves in the trunk of a log and banding the raised sections with iron, produced the forerunner of the Cambridge roll.

In 1815 Edward Poole of Holme Lacy was appointed President. Because of the agricultural depression it was proposed that a petition be sent to Parliament for the abolition of all taxes affecting husbandry.

1816 saw Ben Biddulph as President. The Leominster meeting was abandoned and it was resolved that bulls be exhibited at the Candlemas meeting, sheep and pigs at the summer meeting and heifers at the October meeting. Waning interest was shown throughout 1816 with classes unfilled.

A paper was read by Mr Allen of High Town, Hereford on threshing machines. It is a curiosity in its way, its aim being to prove that labour-saving machinery and threshing machines in particular, were an unmitigated evil, tending to throw honest workmen out of work.

The saga of Mr Broad's 'receipt' for catching rats finally reached a conclusion. He valued it at £1,000 — and this was accepted without demur. The Hereford Society communicated with other societies to raise this amount, but no help was forthcoming, which is perhaps not surprising. However, such was their faith in the device that the Hereford Society raised £500 and the Board of Admiralty stumped up the other £500. A book on its construction was then published of which the relevant details are shown here.

That is the last we hear of Mr Benjamin Broad, who seems to have faded into obscurity, no doubt with a smile on his face.

Mr Price of Foxley was President for 1818 and a subscription was opened for a prize of 20 guineas for the best cultivated farm in Hereford; also prizes for ploughing, the cottager with the greatest number of bee hives and for spinning the largest quantity of flax into thread. When in October the time came round for the ploughing competition, a great quantity of spectators turned up but only one plough team. So they gave the ploughman a gratuity and went home again.

In 1820 it was resolved to present a petition to the House of Commons to exempt agricultural horses from taxation.

The idea was first mooted, in the same year, of holding a regular market at Hereford, for the sale of livestock. Hitherto sales had only been held on fair days.

In 1822 a dreadful depression in every branch of farming produced a thin attendance for the Candlemas meeting. One guinea was given to Anne Cally of Hope under Dinmore for spinning 63 lbs of flax into fine thread, within the preceding year. She made a profit of five shillings a week, towards maintaining herself and four young children, thus rendering unnecessary any appeal for Parish relief.

The serious financial position was discussed and it was resolved to reduce the Society's general meetings to two only, each year. It was also agreed to press for a minimum price of 10s for a Winchester bushel of wheat.

Sir Anthony Lechmere of The Rhydd was appointed President for 1823; he had been a subscriber since 1806. As we shall later see, Sir Anthony was also one of the founders of the Worcester Society. More recently his direct descendant, Sir Berwick Lechmere, was President of the Three Counties Agricultural Society in 1982.

On 21 September 1825 Hereford and Worcester held a joint show, possibly the first attempt at a merger. Hereford sent a considerable quantity of cattle for exhibition where of course they came up against the Shorthorn 'but Hereford carried all before them'.

In the same year occupiers of business premises in the streets of Hereford petitioned to have cattle fairs removed from the streets and asked that they be held outside Bysters Gate (Commercial Road). The farmers objected because the new site offered no shelter — 'and no Inns for transacting business, this being insisted upon'.

After dinner at the Candlemas meeting in 1826 there were discussions on the proposed alteration of the Corn Laws, recent legislation on weights and measures and the banking system of the country. It seems they put more faith in the small private bank than they did in the Bank of England.

Discussions on the Corn Laws continued in 1827, it was feared that the best land would be put down to grass, poorer soils would be neglected and the consumer would be at the mercy of the importer of foreign grain. At the weekly Hereford Corn Market the price of wheat per bushel of 62 lbs was 6s 6d.

At the October meeting it was reported that subscriptions had fallen over the years and at the end of 1826 only thirteen had paid, leaving an adverse balance £144 18s 2d. No prizes had been given and none would be until funds were available.

In 1828 the liabilities became so large and efforts to discharge them so ineffectual that the Society was wound up. The 'New Herefordshire Society' was formed and by this means the old Society was left to wind up its own affairs and the new society commenced unencumbered by debt. All very convenient!

The first meeting of the New Society was held on 19 October 1829, when Broad Street was described as being full of cattle. At dinner afterwards the President, Sir J. Cotterell, believed the Government would do all it could under the heavy weight of National Debt, to assist agriculture. Mr Smythies, in a strong speech, attributed the misfortunes of the country to free trade and blamed Sir Robert Price for supporting these measures. In reply Sir Robert deprecated the return to paper currency. After considerable heated discussion the evening was 'devoted to conviviality', to which the songs of Mr Thomas Cooke, the well known Hereford auctioneer, materially added.

At the Candlemas meeting of 1831 there was a remarkably good show of stock, especially bulls. As the company remained at the festive board until a late hour, it may fairly be presumed that farming prospects were improving. When it is remembered that the dinner commenced at four o'clock in the afternoon and that after almost every toast there was a song, it will be easily understood that midnight still found the party singing, toasting and drinking, the favourite beverage being hot gin and water.

The first toast was of the course 'The Queen', then followed 'The rest of the Royal Family', 'The Chairman Sir Robert Price', 'The Army and Navy', 'The Unsuccessful Candidates', 'The Stewards', 'The Judge', 'The members for the County', 'Mr Turner of Noke', 'The Secretary', 'The Yeomanry', 'The Bishop and Clergy', 'Last Year's Stewards'. All possible toasts having been exhausted recourse was made to sentiment, 'May landlords ever flourish and tenants never fail'.

The Chairman, Sir Robert Price, giving signs of a desire to depart, the health of Lady Price was drunk and the Chairman having replied wound up the proceedings by proposing 'Prosperity to the County'.

He then vacated the chair at what was probably an early hour in the morning and it might have been expected that the rest of the company would follow his example, but not so. The Vice Chairman took the chair and the conviviality was contined until most of those present had attained 'an enviable state of exuberance'. (*Hereford Times*)

In 1832 Broad Street ws not big enough to accommodate all the stock and they extended into St Owen Street, even though they were some distance apart.

The unsettled state of agricultural workers was shown by the case of Henry Williams, sentenced to fourteen years' transportation for sending a threatening letter to Mr Monkhouse of Whitney, because he used a threshing machine.

In October 1833 cattle and sheep were exhibited in St Owen Street and everything else in Broad Street. For the first time a judge was appointed from outside the county, Mr John Buckley of Loughborough.

Shortage of funds necessitated the discontinuance of rewards for large families but two guineas were given to James Oliver of Marden for having twenty-one children.

The Candlemas meeting of 1836 was the first not held in Hereford streets. The show was held in a field of Mr Bosley's, 'near the Kennels without Widemarsh and Byeshead gates'.

At the meeting on 10 August 1836, thanks were given to Messrs Bulmer for offering a meadow for the show; also in that year, a prize was offered for the best plough and ploughman. It was also resolved that the milk allowed for yearling bulls should be restricted to the produce of one cow.

Holding the show in a field found little favour with the public and, after three shows, they moved to a yard owned by Mr Barling of King Street. Here for the first time an entry charge to the public of 1s was made and this raised £13 6s 0d. However, the space was wholly inadequate and they reverted to Mr Bulmer's meadow.

An item from the minutes of 19 May 1844 reads — 'That the secretary write to Mr T. Yeld thanking him for his letter to form a County Herd Book, that at present the committee have determined on nothing more than recommending a very simple form of Herd Book for private use but consider his recommendation well worthy of further consideration'. The proposition was discussed again on 26 June and finally passed on 13 July 1844.

The amount of milk that yearling bulls were allowed to consume seems to have been regularly discussed, it being generally agreed that by the age of nine months they should be fully weaned.

In October 1841 an award for an improved plough was made to Mr R. Griffiths of Devereux Wootton and for a winnowing machine to Peter Walkins of Hereford.

At the Candlemas meeting of 1843 it was agreed that one week's notice of exhibiting must be given; previously it appears that they just turned up on the day. At the same meeting representations were made to the City Council on the inadequate facilities for a market. The scene on market days must have been complete chaos, but it was not until 1854 that a market was finally provided in Newmarket Street.

There was a successful show at Candlemas in 1845, exhibitors coming from as far away as Berkshire.

King's Acre was used for a ploughing match in October 1846 when there three entries in a class for farmer's sons under twenty-five, and eleven entries in the farm-workers' class.

On 1 July 1847 it was recorded — 'That the committee is sensible that the habit of offering premiums for cattle is not without some objection in principle, but it has been found expedient in practice as the best means of drawing farmers together and comparing opinions on agricultural subjects. That in the county of Hereford, perhaps above all others, no agricultural meeting would be popular without it. That attempts are now being made by certain members of the committee to keep down the system of overfeeding and no-one in regular attendance can be insensible of the difficulty of carrying out these sensible reforms. That it must be borne in mind that the primary object of this society at its institution was professedly for the improvement of the breed of Herefordshire cattle, which it has been greatly instrumental in promoting and more still remains to be effected on this point'. (There was a strong section of opinion that Hereford cattle should not be fed any corn).

During the winter of 1849/50 meetings were held in almost every important town in England to consider the question of agricultural distress. The Hereford meeting was called by Mr Barneby, the Sheriff of Hereford, on 6 January 1850.

Farming, as an occupation, was in a more deplorable state than at any time during the century and the responsibility for this was, rightly or wrongly, placed upon Sir Robert Peel for the repeal of the Corn Laws, in an attempt to reduce the price of food.

Believing that the meeting at the Shire Hall was an attempt to restore the old order, it required little effort for the artisans and labourers of Hereford to attempt to prevent it being held.

The meeting was attended by an influential gathering of country gentry, and farmers flocked to it from all parts of Hereford. Well before the time fixed for the meeting to commence the opposition had assembled their forces and, headed by a band, paraded the streets of Hereford carrying two loaves. One was two feet in diameter and depth and labelled 'Free Trade'. The other was 'very diminutive', and labelled 'Protection' and bound with black crepe. The components of the procession, having been regaled with refreshment *en route*, obtained access to the Hall.

As soon as the Sheriff had taken the chair and explained the object of the meeting the Free Traders demonstrated that the purpose of their attendance was to prevent any resolutions being made or any business transacted. In this they were thoroughly successful.

After several vigorous efforts had been made by several speakers to obtain a hearing, without success, it became clear that the attempt to hold the meeting was useless and the Sheriff left the chair. Thereupon there was a stampede of protectionists from the Shire Hall, many of them, especially those who had taken a prominent part on the platform, fearing that they might be the objects of personal violence. In this supposition they were right, for they were followed along the streets, hooted and threatened. Many of them took refuge in shops in High Town and several of the shopkeepers who had afforded them shelter were rewarded by having their windows broken. Not a few of the country people were badly injured and the mob might have proceeded to even greater violence had not the police, who had been instructed not to interfere, at last made their appearance and restored order.

Because of these events and the condition to which the agricultural industry had been reduced, the Candlemas meeting was not a success. The meeting was held on 4 February 1850, the judges Mr C. Brittain of Woodhouse, Mr A. Rogers of Stansbatch and Mr P. Burlton of Lyde. The awards were for yearling bull, Mr Edward Williams, Llowes Court, Radnor; two-year-old bull, Mr T. Roberts of Ivingtonbury; three-year-old bull, Mr W. Racster of Thinghill; aged bull, Mr John Walker of Holmer; cart stallion, Mr Joseph Price of Cross, Kington. Fifteen shillings each were given to Mr Daw of Moreton on Lugg and Mr Preece of Holmer for collections of implements. The latter exhibited a zig-zag harrow, which attracted a lot of attention. Mr P. Turner of Leen took two prizes for roots. Three pounds each were given to James Handley and John Davies for each having brought up eight children. The latter worked for Mrs Davies of Croft Castle for forty years.

In every respect the meeting in October 1850 was in marked contrast to the Candlemas meeting and was an obvious success. There was a capital show of cattle and other stock and a creditable collection of implements.

The judges were Mr Carpenter of Eardisland, Mr E. Longmore of Walford and Mr Boxsidge of Wanstead. In the classes for bull, cow and offspring the judges had great difficulty coming to a decision.

About a hundred sat down to dinner at the Green Dragon Hotel. The President, Lt-Col Clifford, was in the chair and Vice President Mr W. H. Apperley also attended as did the High Sheriff (Mr Cheese) and several prominent landowners.

Mr Yeld of Broome, in replying to the toast to his health, asserted that the cost of keeping an animal from the time it was calved until it was three years old was £18, while the selling price at that time was £12 to £15.

Mr Thomas Bradstock of Penalt made a stirring speech, which promised at one time to create considerable confusion, 'he being called to order by the chairman for trenching on politics', but he was encouraged by the shouts of farmers to 'Go on'. Mr Bradstock urged landlords to fix their rents at a figure commensurate with the price of produce, to fell the hedgerow timber and with the proceeds improve the farm buildings, to destroy game and rabits and to 'abolish those grievous sores — gamekeepers'.

It would seem that the Hereford Society was to a large extent synonymous with the Hereford Herd Book and Breed Society, the one growing out of the other, and a vote of thanks was passed to Mr H. Eyton for the publication of the book. At the same time instruction was given to judges that no preference be shown for variation in colour of the face or body.

The Patroness, Lady Foley, caused some embarrassment by offering a cup for the best cultivated farm in the county. This had to be refused because the accounts were in such a poor state that the Society could not meet the judges' expenses.

Records for the next few years are somewhat sparse but odd bits of information indicate the conditions of the time. Thus we find that no cottager whose rent exceeded six pounds per annum was eligible to compete in the cottager's pig class and no competitor could get first prize for turnips two years running.

Herdsmen in charge of stock were given five shillings for a first prize bull and three shillings for second.

In 1853 it was agreed to restrict the age of indoor servants to thirty, for them to be eligible for an award.

Concern was expressed that Mr Eyton wished to give up the Herd Book because of lack of support. Moving on to 1857 there is a letter from Rev Powell stating that the cost of the Herd Book to his late nephew, W. H. Powell, was thirty pounds. He was however willing to let it go for ten pounds, if a suitable person could be found. Mr Thomas Duckham was requested to take on the publication on condition that he repay the ten pounds and that he publish annually. If he should neglect publication the copyright should be forfeited to the Society and if he wished to discontinue, the Society should have first refusal of the copyright.

In 1859 the Hereford City Council was approached to erect shedding in the market but the Society was told that they could erect shedding at their own expense. Tenders for erection were requested, also for providing refreshment; none were received. This is perhaps not surprising because the balance in hand was only £6 18s 9d.

It was reported that the City Council would not allow refreshments to be sold in the show yard and the Collector of Inland Revenue had made exhaustive enquiries of the Secretary to find out what sort of refreshments were to be sold.

For the 1859 show there were six implement stands entered for the largest collection of improved machinery. There were also six other machinery stands.

In 1860 it was reported that the cost of shedding to the lessee of the market amounted to £500. The Society agreed to pay £25 per annum for the use of the same. The City Council declined to assist in any way.

In 1861 it was agreed that the public could be admitted to the showyard, while judging was in progress on the first day and that they be charged 2s 6d. The admission charge for the second day was to be 2s 6d until noon and 1s afterwards. Bills were posted to the effect that No Smoking be allowed in the showyard.

Some idea of the size of the show can be gleaned from the following for the 1861 event, which was about average: 13 cattle classes with 70 entries, six sheep classes with 10 entries, four pig classes with three entries, four horse classes with 19 entries, one red wheat and one white wheat class, six entries each, and seven entries in the implement class.

In 1865 there was a discussion on whether to hold the show because the Bath and West Show was coming to Hereford that year. It was agreed to carry on after a guarantee fund was opened but in September 1865 it was agreed to cancel the show because a severe outbreak of rinderpest had swept the country. The show was again cancelled, for the same reason, in 1866.

Some idea of the severity of the outbreak can be seen from the following figures. In the week ended 24 February 1866, 17,875 cattle contracted the disease. The Cattle Diseases Prevention Act of 1866 made the slaughter of diseased animals compulsory. The effect was seen at once. In the week ended 3 March 1866, 10,971 were attacked and in the following week 10,056 were killed. At the end of April the weekly tables showed 4,442 attacked, by the end of May 1,687 and in the last week of June 338. In the last week of the year the number had dwindled to eight. In all some 250,000 cattle were affected.

Foot and mouth disease and pleuro pneumonia were endemic at the time and one effect of the slaughter policy was to almost eliminate these two diseases.

Depite this setback the show got going again in 1867. A prize for hops was offered for the first time and classes for breeds other than Hereford cattle were introduced. Even after giving the judges three guineas, a dinner ticket and a bottle of wine, the accounts showed a balance in hand of £155.

In 1869 it was discussed at a general meeting if the usefulness of the Society could be extended by amalgamation with other counties. This was suggested by Mr Trinder, the Gloucestershire Secretary, which was followed in the same vein by a similar letter from Mr Buck of the Worcestershire Society. After much discussion it was resolved — 'That it was not expedient to amalgamate'. Instead it was decided to communicate with the Ross, Ledbury and Leominster Societies. There is no record of the outcome of these approaches.

The relationship that the Society had with Hereford Town Council was not always amicable. At this distance we can only guess at the reasons. Suffice to say that in 1871 the Town Council drew attention to a clause in the lease of the market which forbade its use, presumably for a fair or possibly a sub-let. It was therefore resolved to hold the show in a meadow adjoining the market, occupied by Mr George Partington.

Because of an outbreak of foot and mouth disease in 1872 there was some doubt the show should be held. After members were circulated it was agreed by 68 votes to 33 to hold it. Having got over that hurdle they agreed to send letters to the principal landowners asking for gifts of game for the annual dinner. The letter contained specific instructions as to where and when it should be delivered to the Green Dragon Hotel.

By this time the show had settled into an annual event held in October. There was considerable discussion in 1873 about bringing the date forward and it was resolved that the 1874 show be held on 5 and 6 August. There was also some discussion about making it a three day event.

For this show the lessee of the market offered facilities for £10 — which was accepted. For the 1873 show a poultry section was added and in 1874 a horticultural section. £100 was given in prize money for the poultry section.

In addition Mr Partington's meadow was again hired, a grandstand 100ft long was built and jumping competitions introduced. This was also the first time that a military band appeared. This seventy seventh anniversary was reckoned to be 'in importance very far in advance of any previous meeting of the Society'.

It would seem, however, that pride went before a fall, because the receipts for the year were only £1,336 9s 10d against expenditure of £1,659 18s 5d, leaving an adverse balance of £323 7s 7d. The guarantors were called upon to pay 10s 6d in the pound, in order to save the Society from bankruptcy.

Because of this calamity negotiations were entered into with the Shropshire and West Midland Society, with a view to amalgamation. However, it was reported on 17 February 1876 that an appeal for cash had been successful and it was decided that amalgamation was not desirable. Rev Sir George Cornwall and Mr Thomas Duckham represented the Society.

At that time there was another attempt to amalgamate with Worcester, the sub-committee appointed being Lord Bateman, Rev Sir George Cornwall, H. Taylor, H. M. Edwards and H. Haywood. All these negotiations were carried on in a somewhat half-hearted manner, precipitated by a shortage of cash. As soon as the financial situation was resolved all thoughts of amalgamation ceased.

Then the character of the show started to change. Until then it has been purely a cattle show. With the introduction of ring events it started to move into the format that we know today.

It was further decided to hold the show every third year at Hereford and in the intervening years at one or other of the principal towns in the county, on the understanding that £250 be raised locally towards the cost.

The 1877 show was held in Hereford and it was agreed that in future the show would be held on the Tuesday, Wednesday and Thursday following the Royal. It was also agreed that in future the Mayors of the towns visited should be members of the committee.

Mr Thomas Duckham resigned as Secretary of the Society and as Editor of the Herd Book, but no successor could be found. Finances were so bad that they were unable to exercise their option to purchase the copyright.

Ledbury and Kington were proposed as venues for the 1879 show and there was a large majority in favour of Kington. About twelve acres were required.

In 1880 there was no invitation to hold the show in the county, so it was decided to extend it beyond the county boundary and Hay on Wye was chosen. There was a section for old implements and birth dates of entries were required for the catalogue.

The 1881 show at Leominster incurred a loss of £120 11s 3d, which was put down to the trade depression. Mr T. Duckham resigned as Secretary for the second time and he was succeeded by Mr Alfred Edwards, who had been secretary of the horticultural section. Mr Edwards' appointment was the beginning of a period of family service by father and son, which lasted until Mr T. H. Edwards retired in 1946 and Mr Glynne Hastings was appointed. Their joint contribution to the Societies was immense and cannot easily be measured.

The charges made by the railway companies for the carriage of stock and material seemed to be a regular topic of discussion and much time was spent sending deputations to meet the railway authorities.

At a meeting of the prize list committee in February 1882 it was agreed that all Hereford cattle exhibited should be entered in the Hereford Herd Book.

The country suffered another outbreak of foot and mouth disease in 1883. On 30 May a deputation visited Hereford magistrates to get their consent for the return of the show stock from Abergavenny, where the show was to be held. This was granted but the deputation also had to visit Abergavenny magistrates to get their consent.

In 1884 a serious attempts to extend the show were made. Approaches were made to Radnor, Brecon, Monmouth and Glamorgan, with a view to amalgamation with one of these county societies. None of them were keen so Worcester and Gloucester were approached. Worcester was quite interested but Gloucester was not. They then once more tried the Shropshire and West Midland. Negotiations seemd to be going quite well but, when it came to the vote, there were seven in favour and eight against, so the negotiations fell through.

Complaints occupied a lot of time at stock meetings. There was a tremendous row in 1890, with a great splash in the *Hereford Times*, because a colt that got first prize at the Royal failed to get placed at Hereford. Accusations were made that he had been put down in favour of a colt owned by someone from whom the Society hoped to obtain a large subscription. However, Wyndham Briar, the Society's vet, pronounced the colt unsound.

Complainants were frequently represented by solicitors at these hearings and the discussions must have been pretty tense.

In 1891 the Hereford Society supported and went to great lengths to start a Dairy School in the county. Classes in butter-making were held at various centres. The courses were of ten days' duration, certificates of proficiency being awarded to the participants on successful completion. The scheme was funded by the County Council but most of the organisation was done by local committees.

The Society visited various towns by invitation and it was a condition of acceptance that the town find the sum of £250 and provide a free water supply. The show was growing and it was becoming difficult to find a sufficient number of towns in the county, or not too far away, that could accommodate it.

Because of the this difficulty in 1894 Worcester City were asked if they could accommodate the show.

GENERAL RULES AND ORDERS.

A President, Deputy President, Committees, Treasurer and Secretary, with usual powers, are annually elected by a general ballot, in order to assist in carrying into effect the designs of the Society.

Annual Subscriptions of not less than One Guinea each, to commence from the first of January in each year, entitle a person to be a Member; but any smaller sums are received, and give the privilege of being present at the Meetings, but not that of voting on any question before the Society; the names of all Subscribers are published, and any persons who have given in, or may give in, their names as Subscribers, are deemed indebted to the Society in the amount of their several subscriptions, until they give notice in writing to the Secretary of their intentions to withdraw them. Breeders, not being inhabitants of the county, are admitted to shew stock, for the Society's Premiums, on becoming Subscribers

The Society meet four times in the year, viz. on the first Monday in March, in Hereford; on the first Monday in June, in Hereford; on the   of   at Leominster, and on the day before the October Fair in Hereford, which October Meeting is deemed the Anniversary of the Society, when new Officers are appointed; the Rules of the Society revised; former Premiums awarded, and new ones proposed.

The Meetings are held alternately at the City-Arms, New Inn, and Green Dragon, in the city of Hereford, and the Chair taken at twelve o'clock precisely; the Leominster meeting is held at the Crown Inn.

All propositions are laid before the Society, at one Meeting, and decided on at the next following.

Any three Members are competent to act as a Committee in promoting the objects of the Society, in the intervals between the General Meetings; and all Committees are open to a Subscriber of one Guinea per annum.

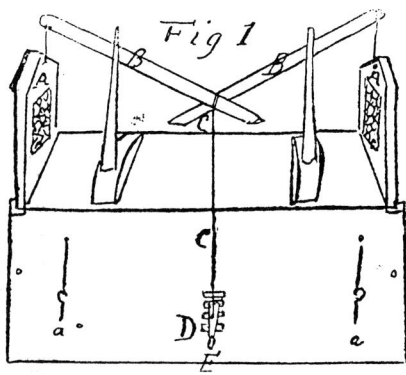

o William Downes, of Hinton, second best Crop of Turnips, *twice hoed*, Silver Plate - - - - 8 3 0

o T. A. Knight, Esq. for producing the best new variety of the Apple, raised from seed, a Silver Goblet - 6 6 0
N. B. This fruit was directed to be called the GRANGE APPLE, and was raised under Mr. Knight's immediate inspection, from the kernel of the apple called the *Woodcock*.

o John Evans, of Peterchurch, for bringing up Thirteen Children, without assistance from the parish - - - - - 3 3 0

o Samuel Suff, of Hom Lacy, for bringing up Eleven Children, as above - - - 2 2 0

o John Bevan, of Dindor, for bringing up Ten Children, as above - - - - 2 2 0

John Wall, for Forty-three years' faithful Service with Mr. Apperley, of Withington - 3 3 0

Michael Dukes, for Thirty-six years' Service with Mrs. Jordan, of Bradfield - - - 2 2 0

o John Guy, for Thirty-four years' Service with Mr. and Mrs. Caldwall, of Hom Lacy - - 1 1 0

PREMIUMS WERE ADJUDGED

o T. A. Knight, Esq. for Ploughing the greatest number of Acres, with Oxen, *worked singly*, Silver Plate - - - - 5 5 0

To Mr. William Watkins, of Brinsop, for having the best Crop of Turnips, *twice hoed*, a Silver Goblet - - - - - 5 5 0

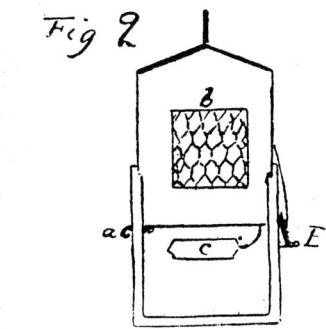

Fig. 1 in the above plate represents the trap when set.
A A, are the falling doors with the wire grates.
B B, the levers which support the door.
C C, the small cord which is attached to the trigger.
D, the trigger. A small thin plate of iron, two inches long and about six lines broad, with a hole in the middle to receive the cord or chain which passes the levers.
E, the extremity of the neck of the bridge.
a, The pin which holds up the door

Fig 2. This section of the trap gives its appearance end ways, and is intended to show the position of the door, bridge, trigger, &c, as when set for catching, except that the pin is in its actual state, and the bridge must be covered with straw, agreeably to the directions hereinafter given.
E, The end of the neck of the bridge projecting two-thirds, or not more than three-quarters, of an inch.
a, The pin which supports the door.
b, The wire grate or window.
c, The bridge hanging horizontally half an inch above the floor.
In setting the traps care must be taken that the triggers have not too deep or strong hold in the neck of the bridge; and to prevent the probability of this happening the depth of the notches in the neck of the bridge should not be more than what will be necessary to cause the bridge to be supported till the rats tread upon it. The trigger should also stand very nearly perpendicularly, as the

ABOVE: Mr Bertram Bulmer, into whose meadow the Hereford Society overflowed. LEFT: Results of Hereford Agricultural Show held 22 October 1799. OPPOSITE: Rules of the original Hereford Society founded in 1797. RIGHT: Plans for Mr Broad's rat trap, for which the Society paid him £1,000.

Herefordshire
New Agricultural Society, as
remodelled from the old Society
A.D. 1829

Regulations –

Subscribers of one Guinea each per ann
to be members, & to be considered as such
until written notice to the contrary, be
delivered to Mr John Bell, who is appointed Collector of the Funds –

Premiums presented to the Society to
be subject to the same regulations as
all other premiums.

None but members whose names appear on the list of subscribers before
the days of meeting, are qualified to
receive a premium –

The same premium not to be awarded
during two successive years, to the same
person.

The successful Candidates may receive
their premiums in money or plate
at their own option.

Gentlemen /–

I am instructed by the Corporation
to inform you in reply to your Memorial dated the
28th of May 1856 and addressed to the Mayor and
Town Council of Hereford that those Gentlemen have
agreed that your Society shall have the free use of the
New Cattle Markets on Saturday the 18th day of Octr
next on your "Committee guaranteeing that any Animals
sold in the yard shall pay the same toll as if sold in
the Markets" as proposed in your Memorial

I remain Gentlemen
Your very obedt Servt
Rich d Johnson

Extract from the minute book of 'The New Herefordshire Agricultural Society'. RIGHT: Medal awarded to Mr J. Davis of Webton Court, for Long Wool sheep in 1859. (Courtesy Hereford Museum) BELOW: Letter giving permission to use the New Cattle Market.

LEFT: Hereford Bull of around 1860. Even allowing for artistic licence, the conformation of the breed has altered in more modern times. RIGHT: Shorthorn Cow of the mid-eighteen hundreds; large areas of patchy fat may be discerned. What was then a desirable feature is now regarded as waste. BELOW: Extract from Hereford Minute Book for 1847.

HEREFORDSHIRE AGRICULTURAL SOCIETY.

President for 1858:—ROBERT BIDDULPH, Esq.   Vice-Presidents:—Mr. J. DAVIES, Webton Court, Mr. G. YELD, Twyford.
Stewards:—Mr. W. TAYLOR, Thingehill, Mr. P. TURNER, The Leen.
Treasurer and Secretary:—Mr. J. T. OWEN FOWLER, Hereford.

THE THIRTY-FIRST ANNUAL SHOW

Will be held in the NEW CATTLE MARKETS, HEREFORD, on TUESDAY, the 19th OCTOBER, 1858.

When the following **PREMIUMS** will be offered for Competition:

CATTLE.

	£	s.
CLASS 1.—For the best Bull, Cow, and Offspring; the Offspring to be bred by the Exhibitor, and to be calved on or after 1st July, 1857.		
First Prize (gift of the Citizens of Hereford)	25	0
Second Prize	10	0
Third Prize	5	0

Four at least to be shown, or the premiums will be withheld.

	£	s.
CLASS 2.—For the best Bull, calved on or after the 1st of July, 1857 (gift of the citizens of Hereford)	20	0
Second in this class	10	0
Third ditto	5	0
CLASS 3.—For the best Bull, calved on and after the 1st of July, 1856	6	0
Second in this class	3	0
CLASS 4.—For the best Bull, calved previous to the 1st of July, 1856 (premium gift of Rev. W. T. K. Davies.)	5	5
Second in this class	3	0
CLASS 5.—To the Tenant-farmer, being a subscriber, who shall exhibit the best lot of Beasts, irrespective of sex, bred by himself, and fed without corn or cake, under two years and six months old, in proportion to the quantity of land that he occupies, as follows (premium gift of the Lord Bateman)	5	5

The Tenant-occupier of 100 acres, two beasts.
" " " 200 acres, four beasts.
" " " 500 acres, six beasts.

	£	s.
CLASS 6.—For the best pair of Heifers, calved on or after 1st of July, 1857 (premium of G. Clive, Esq., M.P.)	5	5
Second in this class	3	0
CLASS 7.—For the best pair of Heifers, calved on or after the 1st of July, 1856 (premium gift of Sir H. G. Cotterell, Bart., M.P.)	5	5
Second in this class	3	0
CLASS 8.—For the best pair of Steers, calved on or after the 1st of July, 1857 (premium gift of J. King King, Esq., M.P.)	5	5
Second in this class	3	0
CLASS 9.—For the best pair of Steers, calved on or after the 1st of July, 1856 (premium gift of T. W. Booker Blakemore, Esq., M.P.)	5	5
Second in this class	3	0
CLASS 10.—For the best pair of Steers, calved on or after the 1st of July, 1855 (premium gift of Lieut.-Col. Clifford, M.P.)	5	5
Second in this class	3	0
CLASS 11.—For the best four Steers, calved on or after the 1st of July, 1855, bred by one person, and to be the property of the Exhibitor at the Hereford May Fair, 1858 (premium gift of Sir Velters Cornewall, Bart.)	5	5

• This premium is open to all subscribers, or to non-subscribers upon payment of 10s. 6d. entrance.

	£	s.
CLASS 12.—For the best lot of Breeding Cows or Heifers, not under three years old, that have had a calf within six months, or shall be in calf at the time of showing (premium gift of the Citizens of Hereford)	20	0
Second in this class (gift of Mr. John Ford, jun.)	5	0

The Occupier, if not exceeding 100 acres, to show two beasts.
" " 150 acres, to show three beasts.
" " 200 acres, to show four beasts.
And in the same proportion for every additional 50 acres.

	£	s.
CLASS 13.—For the best Fat Cow, of any age (premium gift of F. R. Wegg-Prosser, Esq.)	5	5

SHEEP.

	£	s.
CLASS 14.—For the best pen of twenty Breeding Ewes, of any breed (gift of W. P. Herrick, Esq.)	5	5
CLASS 15.—For the best pen of four Yearling Wethers, long-wool, premium	3	0
CLASS 16.—Ditto ditto Ewes, ditto premium	3	0
CLASS 17.—Ditto ditto Wethers, short-wool, cross-breeds not excluded, premium	3	0
CLASS 18.—Ditto ditto Ewes, ditto ditto premium	3	0

PIGS.

	£	s.
CLASS 19.—For the best Boar Pig, under two years old, premium	3	0
CLASS 20.—For the best Breeding Sow that has brought a litter of pigs within four months of the date of showing, or, being in pig, shall produce a litter on or before the 19th February, 1859, premium	2	0
CLASS 21*.—For the best Cottager's Pig, premium	1	0
Second in this class	0	10

* *The competitors for this premium must send in a written notice and recommendation from a member of the Society, to the Secretary, on or before the 18th day of September, 1858.*
No Cottager to be eligible to compete who is the occupier of more than half an acre of land, or whose rent exceeds £6 per annum.
Each unsuccessful competitor to be allowed 1s. for the loss of his day's labour from the Society's fund.

EXTRA STOCK.

	£	s.
CLASS 22.—A sum not exceeding £5 will be placed at the discretion of the Judges to be awarded for Extra Stock, the property of subscribers	5	0

HORSES.

	£	s.
CLASS 23.—For the best Cart Stallion (premium gift of H. Lee Warner, jun., Esq.	5	5

This premium is NOT limited to animals bred by the Exhibitor, and is open to public competition on payment, if not a subscriber, of one guinea entrance. It will not be presented until the October Meeting, 1859, as the winner must engage that the Horse exhibited by him, shall, in the ensuing season, cover regularly in the county of Hereford.

	£	s.
CLASS 24.—For the best Cart Mare and Foal at foot. Gift of Sir E. F. S. Stanhope, Bart.)	5	5
CLASS 25.—For the best three-years-old Colt, Gelding, or Filly, suited for hunting purposes, to have been bred by the Exhibitor, or in his possession ten months prior to the day of show. Age to be reckoned from 1st January. Gift of Mr. W. James	5	0

WHEAT.

	£	s.
CLASS 26.—For the best Sample of Seed Wheat, grown by a Tenant-farmer in the county of Hereford, farming not less than 200 acres; the samples not to be less than a sack of five imperial bushels (premium gift of the Right Hon. the Lady Emily Foley)		

IMPLEMENTS.

	£	s.
CLASS 27.—For the best and largest collection of Improved Agricultural Machinery and Implements (premium gift of the late John Arkwright, Esq.)	5	5
Second best (by the Society)	2	0

A list and description of the Implements, &c., exhibited to be delivered to the Secretary on or before the 18th September, 1858.
This premium is open to all subscribers, or non-subscribers, upon payment of £1 1s. entrance.

REGULATIONS.

Persons intending to exhibit Stock must send notice in writing to the Secretary, who will furnish them with printed forms of Certificate, which must be duly filled up, signed, and returned to the Secretary, on or before *Saturday, 18th September*, 1858, after which date no entry will be received; with the exception of Extra Stock, for which entries may be made up to Saturday, 9th October, 1858.

A charge of Five Shillings for each lot entered and not sent for exhibition will be made in every Class (excepting extra stock), unless the Committee approve of the reasons given for such non-exhibition.

The Secretary and Stewards will attend at the Committee Room, at the house adjoining the Cattle Markets, Hereford, from 7 a.m. to 12.30 noon, on Monday, the 18th October, when a Ticket will be given to each exhibitor to affix on or near the animal shown, which ticket will correspond with one fixed in the proper class in which the animal is entered for exhibition, and no animal will be admitted into the yard without such ticket.

The Show Yard will be closed punctually at 12.30 noon, on Monday, 18th October, 1858, after which time no Animal or Article for exhibition will be admitted, with the exception of Extra Stock, which will be admitted in the Show Yard up to 7 o'clock on Tuesday

Schedule of classes for Hereford Show in 1858.

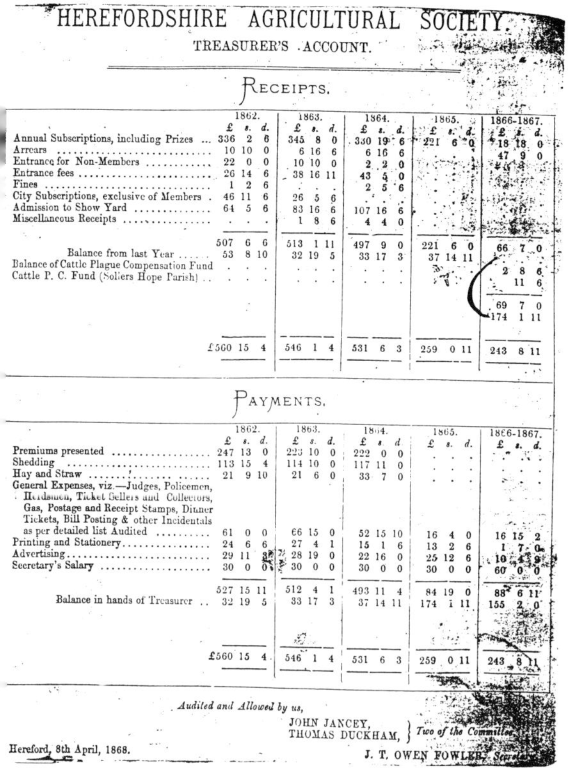

HEREFORDSHIRE AGRICULTURAL SOCIETY.

TREASURER'S ACCOUNT.

RECEIPTS.

	1862. £ s. d.	1863. £ s. d.	1864. £ s. d.	1865. £ s. d.	1866-1867. £ s. d.
Annual Subscriptions, including Prizes ...	336 2 6	345 8 0	330 19 6	221 6 0	18 18 0
Arrears	10 10 0	6 16 6	6 16 6	. . .	47 9 0
Entrance for Non-Members	22 0 0	10 10 0	2 2 0	. . .	[illegible]
Entrance fees	26 14 6	38 16 11	43 5 0	. . .	. . .
Fines	1 2 6	. . .	2 5 6	. . .	. . .
City Subscriptions, exclusive of Members .	46 11 6	26 5 6	. . .	. . .	. . .
Admission to Show Yard	64 5 6	83 16 6	107 16 6	. . .	. . .
Miscellaneous Receipts	. . .	1 8 6	4 4 0	. . .	. . .
	507 6 6	513 1 11	497 9 0	221 6 0	66 7 0
Balance from last Year	53 8 10	32 19 5	33 17 3	37 14 11	
Balance of Cattle Plague Compensation Fund	. . .	. . .	. . .		2 8 6
Cattle P. C. Fund (Soliers Hope Parish) ..	. . .	. . .	. . .		11 6
					69 7 0
					174 1 11
	£560 15 4	546 1 4	531 6 3	259 0 11	243 8 11

PAYMENTS.

	1862. £ s. d.	1863. £ s. d.	1864. £ s. d.	1865. £ s. d.	1866-1867. £ s. d.
Premiums presented	247 13 0	223 10 0	222 0 0	. . .	. . .
Shedding	113 15 4	114 10 0	117 11 0	. . .	. . .
Hay and Straw	21 9 10	21 6 0	33 7 0	. . .	. . .
General Expenses, viz.—Judges, Policemen, Herdsmen, Ticket Sellers and Collectors, Gas, Postage and Receipt Stamps, Dinner Tickets, Bill Posting & other Incidentals as per detailed list Audited	61 0 0	66 15 0	52 15 10	16 4 0	16 15 2
Printing and Stationery..............	24 6 6	27 4 1	15 1 6	13 2 6	1 7 0
Advertising....................	29 11 3	28 19 0	22 16 0	25 12 6	10 4 9
Secretary's Salary	30 0 0	30 0 0	30 0 0	30 0 0	60 0 0
	527 15 11	512 4 1	493 11 4	84 19 0	88 6 11
Balance in hands of Treasurer ..	32 19 5	33 17 3	37 14 11	174 1 11	155 2 0
	£560 15 4	546 1 4	531 6 3	259 0 11	243 8 11

Audited and Allowed by us,

JOHN JANCEY,
THOMAS DUCKHAM, } *Two of the Committee*

Hereford, 8th April, 1868.

J. T. OWEN FOWLER, *Secretary*

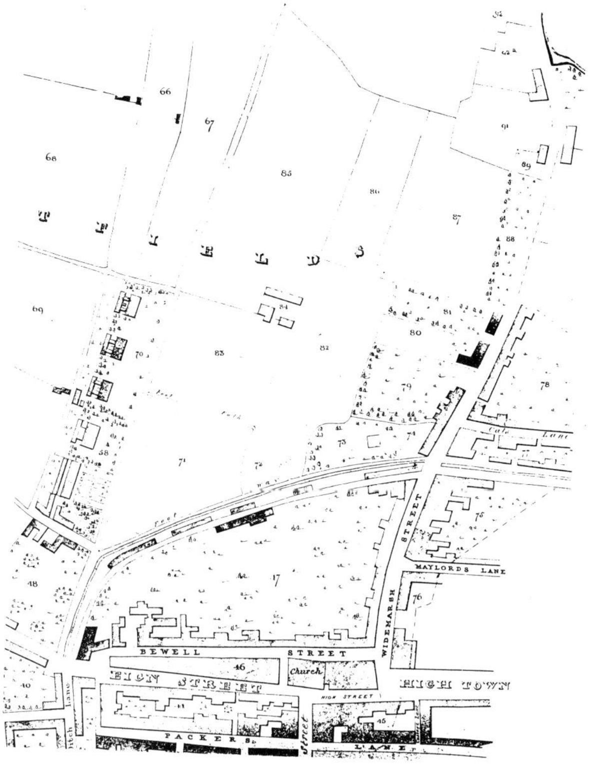

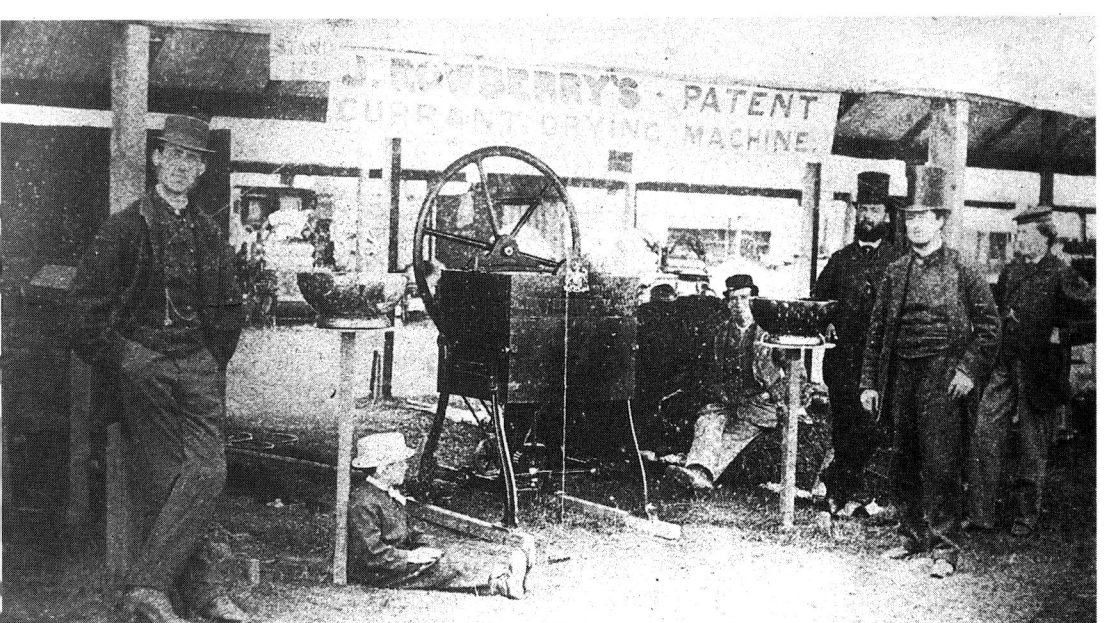

LEFT: Statements of Accounts for the years 1862-1867. RIGHT: Map of Hereford c1830. The fields 71, 83 and 85 were owned by Mr Bulmer. Mr Bosley also owned land in the same area. Plot 82 is the approximate site of the 'New Cattle Market'. BELOW: A stand at Hereford Show in 1865.

WORCESTERSHIRE AGRICULTURAL SOCIETY.

JOHN COUCHER DENT, ESQ., PRESIDENT.

JUDGES.

CATTLE, SHEEP, AND PIGS.—Messrs. S. Bloxsidge, H. Chamberlain, and — Stratton.
HORSES.—Messrs. R. Bloxsidge, T. Potter, and T. Boughton.
IMPLEMENTS.—Messrs. J. Jackson, R. Guilding, and B. Hall.
ROOTS.—Messrs. Chadwick, Moore, and Buchan.
WOOL.—Mr. James Coucher.
CORN.—Messrs. Lacy, Goodwin, and J. Guilding.
HOPS.—Messrs. Jno. Tyler and Jno. Palmer.
FRUIT.—Messrs. Haywood, Smith, and Onley.

The annual general meeting of this society, and exhibition of stock, implements, corn, roots, &c., took place to-day (Friday), in the Cattle Market. The weather was unfavourable in the early part of the morning, but it cleared up some time previous to the opening of the show yard. The show was on the whole a large one; there being a better entry than usual, and an increased competition in all the classes. In the cattle department, Mr. Henry Hill took Viscount Elmley's prize for the best bull two years old and upwards, and he also succeeded in obtaining the prize for another bull under two years. The animals shown by Mr. B. Hall, and Mr. J. S. Walker in the same class were highly meritorious. In class 3 (short-horns and other breeds, not Herefords) Earl Beauchamp gained the prize offered by himself, for the best cow in milk, bred and reared by the exhibitor; and Mr. W. Woodward was awarded the society's prize, for the best bull, two years old and upwards. Mr. J. S. Walker, Mr. E. B. Guest, Mr. T. Harris, and Mr. B. Hall, also obtained prizes. The show of sheep was very good, Messrs. Moore, Randell, J. Lett, Adcock, T. B. Brown, and Trinder, being among the principal exhibitors. Mr. J. Lett gained the prize for the best ten breeding ewes, long-woolled, and Mr. W. Moore that for ten best ditto, short-woolled. The well-known reputation of these breeders will convey an ample idea of the quality of the stock shown, which was first-rate.

Pigs were numerous and of excellent description. Mr. H. Hudson was deservedly awarded a prize for a breeding sow and eight pigs shown by him, and also took one for the three best open sows of the same litter. Mr. H. Hill's young boar was a very superior animal, and its merit, it will be seen, was duly appreciated. The cottagers came out strongly for their prizes, and exhibited some very good animals. The prize pigs in this class were remarkable for their size and condition.

Of horses, the show, though not a very large one, comprised some animals of a very useful character, and the competition was close.

The department of roots was very plentifully filled. Mr. Humpidge, Mr. John Ellis, and Mr. G. McCann showed swedes and mangold wurtzel, and Mr. B. Hall some mangold. Mr. R. Guilding had a capital collection as extra stock, comprising twenty samples each of long red and yellow mangold wurtzel and swedes, sixty of turnips, of three different varieties. He obtained the society's premium for twelve best common turnips.

Of fruit the show was very meagre, of course attributable to the unfavourable season.

Excellent samples of white and red wheat, and also of barley were exhibited. There was some wool of good quality, and also a few hops pitched.

The implement yard was more than ordinarily furnished. Messrs. Bell and Hall, Matthews Brothers, Horton and Griffiths, and Hunter and Co. (Worcester), Messrs. Mapplebeck and Lowe (Birmingham), Mr. E. Grubb, jun., (Bromyard), Mr. Simpson, (Evesham), and Mr. John Burrow (Leigh), severally contributed. The exhibition was mainly noticeable from the number and variety of useful implements it contained. Messrs. Matthews showed several of Samuelson's improved turnip cutters and slicers, Richmond and Chandler's chaff cutters, Bentall's patent broadshare ploughs, Phillips's poppy extirpator, G. O. ploughs, with wheels and skim; a bean splitting mill, and Wedlake's patent haymaking machine. In Messrs. Bell and Hall's collection was an improved iron and ridge plough, of their own manufacture, improved winnowing machines, chaff engines, with horse power; turnip drills and weighing machines; also Williams' new patent digging machine, manufactured by themselves; an excellent implement, calculated to do its work well and efficiently. Messrs. Horton and Griffiths exhibited some rick, waggon, and cart covers. Mr. John Burrow, (Leigh,) a six-horse power steam engine and winnowing machine. Messrs. Hunter & Co. exhibited a four-horse grubber, a skim plough and grubber combined, and a G.O. plough, fitted with wheels and skim coulter. Mr. Simpson, a furrow drill for general purposes, and two and three-furrow wheat drills, Bentall's prize pulping machine, a set of whippletrees and a registered chaff engine. Mr. E. Grubb, jun., (Bromyard,) had in his collection a new mowing machine, by Shanks & Son, Arbroath, Forfarshire; Moore's new London washing machine; two combined winnowing and blowing machines, improved hay-making machine, combined threshing machines, (by Messrs. Humphries,) two improved portable steam engines, (by Clayton & Co., Lincoln,) and a miscellaneous assortment of ploughs, chaff machines, scuffles, drills, &c. Messrs. Mapplebeck and Lowe showed some of Howard's and Bentall's patent ploughs, Coleman's cultivator, a prize chaff-cutter of their own manufacture, a platform and sack weighing machine, Gardner's patent turnip cutter, &c. The steam machinery was exhibited in motion.

Among the company in the show-yard and who also visited the poultry exhibition, were Earl Beauchamp, Viscount Elmley, M.P.; J. H. H. Foley, Esq., M.P.; Capt. Candler, Col. Clowes, Col. Scott, the Mayor of Worcester (John Goodwin, Esq.), Rev. R. Forester, Messrs. J. G. Watkins, T. C. Hornyold, F. and W. Woodward, T. B. Browne (Gloucestershire), H. Lakin, C. Randell, R. Ashwin, J. Lett, Twinberrow, A. Pike (Longdon,) J. Coucher, J. S. Smith, W. S. P. Hughes, Tolley, Herbert, Dale (Spetchley), T. Pitt (Wolverley), T. Quant, &c., and most of the exhibitors.

THE PLOUGHING AND DIGGING

took place at Mr. W. Coney's, on the Middle Battenhall

LEFT: Part of the extensive report of the 1856 Worcester Show in *Berrows Journal*. RIGHT: In the early days of showing, pigs were often so fat that they could not get up. That is noticeable in this example. BELOW: The central detail of the solid silver salver presented to John Crane Nott of Hallow, Secretary of the Worcestershire Agricultural Society 1838-1843. It was largely due to his enthusiasm that the Society was revived in 1838.

32

PLOUGHING AND PEARS —
The Worcestershire Agricultural Society

Early accounts of the Worcestershire countryside speak of a cold, heavy land that is difficult to work. You get the impression of peasants eking out a miserable existence on unrewarding soil in the company of nondescript cattle of the Shorthorn type.

Writers do, however, admit of more fertile soils in the Evesham-Pershore area and they also acknowlege that Worcestershire is a good place for growing pears. Varieties mentioned are Bishop's Thumb, Swan's Egg, Glou Morceau, Brown Beurre, Marie Louise, Jargonelle, Barland, Taynton Squash, Red Horse, Hampton Rough, Longland, Sanguinole and, the most famous of all, the Black Pear of Worcester, which of course appears on the Worcester city coat of arms.

Some of the more progressive Worcester men supported the Hereford Agricultural Society, among the earliest members, Hon E. Foley MP for Worcester and Anthony Lechmere of The Rhydd.

At a meeting held on 29 January 1816 at the Hop Pole, Worcester, chairman Viscount Elmley, supported by Lord Foley, Sir T. E. Winnington and Sir W. Smith, proposed the formation of a Society to be called 'The Worcestershire Society for the Encouragement of Arts, Manufactures and Commerce and for the promotion of Industry and Emulation among Servants and Labourers'. The proposition was seconded by Lord Foley and passed unanimously. About a hundred subscribers put down their names.

A committee was formed and met at the Guild Hall coffee house on 23 February, with Lord Foley in the chair. A prize list was drawn up for the first show, which was held on 14 June 1816 at the Star and Garter Inn. It was as follows:

Class 1	The best three year old heifer in milk	£5
Class 2	The best four year old heifer in milk	£5
Class 3	The best fine woolled shearling ram of pure Ryeland or Southdown breed	£5
Class 4	The best fine woolled ram of pure Ryeland or Southdown breed any age	£5
Class 5	The best long woolled shearling ram	£5
Class 6	The best long woolled any age	£5
Class 7	The best pen of 3, two shear short wool wethers	£5
Class 8	The best pen of 3, long wool wethers	£5
Class 9	The best cart stallion	£10

Awards were Class 1, 5, 6 and 8 Thomas Moore, Cofton Hall; Class 2 James Price, Little Malvern; Class 3 No Competitor; Class 4 and 7 Edmund Welles, Earles Croome; Class 9 Thomas Saunders, Feckenham Lodge.

The Society held its second show on 4 October that year and gave prizes for heifers, pair of steers, a boar, a sow, new implements, best method of destroying turnip fly, a seedling apple, a seedling pear, long service men and women and the labourers who had reared the largest families without Parish relief.

In 1817 the Society adopted the less cumbersome title of 'Worcestershire Agricultural Society', and the enlarged schedule included classes for bulls, the best crop of not less than five acres of turnips and ploughing half an acre with oxen or horses, first and second prizes given to the ploughmen.

Two shows were held each year 'in Mr Hope's field at the top of the Tything adjoinig the Shrubbery'. Dinners were held alternately at the Star and Garter and the Hop Pole.

In May 1819 George Bentley, Auctioneer, succeeded Jeremiah Herbert as Secretary. It is clear that, in common with other agricultural societies, finances were not all that sound because a note in 1820 reads 'We are sorry that the Society is in much disadvantage because of Subscribers neglecting to pay up their subscriptions'. Slow though they may have been in paying up they very soon hit on the idea of running a sweepstake on the winners. In 1819 John Price won fifteen guineas on a pair of heifers and sixteen guineas on a pen of breeding ewes.

Another note reads 'we are glad to hear that many farmers in the county are raising seedling pear trees. If this practice is generally adopted, Worcester will regain its celebrity for this type of produce. It is the opinion of many well informed persons that good fruit cannot be raised by any other means'.

Mr Price, replying to the drinking of his health as successful prize-wiiner, said that he did not approve of judges being appointed from among their number. They should come from outside the County.

The Society continued to hold two annual shows in Mr Hope's field, but nothing remarkable is recorded. They had a single show in September 1826 after which the Society became defunct.

In 1830 Rev Henry Berry of Acton Beauchamp attempted to revive it. Although sixty members, who enrolled at the Crown Hotel, promised their support, the effort proved fruitless.

The Society was in fact revived in 1838 and arranged a ploughing match in a field belonging to John Crane Nott of Hallow. This was followed by a dinner attended by the Town Council. Speed in ploughing the measured quarter acre was an important factor in this competion; results were:

Long Teams		G.O. Ploughs	
Mr Nott's	55 minutes	Mr R. Spooners	76 minutes
Mr Walker's	57 minutes	Mr B. Forresters	79 minutes
Mr Hooke's	59 minutes	Mr P. V. Onslow's	83 minutes
Mr Onslow's	61 minutes	Mr Walker's Oxen	81 minutes
Exors Mr Jones	62 minutes		
Lord Beauchamp's	63 minutes		
Mr R. Spooner's	89 minutes		

Prizes were also awarded for 'long and faithful service' and for bringing up the largest family without the aid of Parish relief. First prize for the latter was given to John Brazier of Hanbury who had brought up fifteen children. He received £2 and the chairman expressed the wish that this would enable him to purchase some comforts for the coming winter.

At the dinner Mr Spooner announced that General Lygon was offering a prize of £5 for the best bull, thus starting the cattle classes in the revived Society.

The Mayor, Mr George Allies, in reply to a toast, commented that the Council 'has not done at all too much in building a market' and was happy that this had been done during his Mayoralty. A Mr Deighton, speaking on the same subject, expressed the wish that the cost would soon be paid off and there would be no necessity for tolls.

The 1839 show saw the introduction of two bull classes, both won by Herefords and two classes for sheep. The other event of note was the appearance of Robert Tombs, an 85-year-old, who had been in the service of Mr W. Smith of Crowle for 53 years. Presenting him with £5 the chairman, the Earl of Coventry, said that he considered him to be the Methuselah of goodness, that he hoped the rest of his life would be spent in the same manner and the result would be a happy and glorious death!

Ploughing matches were held in 1839 and 1840 on Mr E. Herbert's farm at Powick. These were succeeded in 1841 by a comprehensive show held in the New Cattle Market, Worcester. There were classes for cattle, sheep, pigs, horses, swedes, carrots, mangolds, seedling apples and pears, and there was also a machinery section.

On 1 October 1843 at the Crown Hotel, Worcester, the Society held a dinner and presented John Crane Nott with a silver salver for his services as Secretary since its reconstitution. The salver, 26 inches in diameter, was of solid silver and weighed nearly 200 ounces. It was beautifully chased with episodes of four agricultural scenes and suitably inscribed in the centre. Around the circumference on the obverse side were engraved the names of 195 subscribers.

The salver came into the possession of E. V. V. Wheeler of Newham Court and, on its rediscovery in 1934, Alderman R. B. Worth, Lickhill Manor, Stourport purchased it and presented it to the Worcestershire County Council to be preserved in perpetuity in the County Hall.

The principal exhibitors between 1840 and 1850 were:

Horses	Cattle
Thomas Saunders, Hanbury	E. Larking, Beauchamp Court
Thomas Baldwin, Barnt Green	Earl Beauchamp, Madresfield Court
E. Larkin, Beauchamp Court	H Joseph Moore, Stanford
E. Herbert, Powick	N. Smith, Horsham
J. G. Watkins, Ombersley	E. Herbert, Powick
W. Rodgers, Martley	John Herbert, Powick
C. Hudson, Wick, Pershore	H John Price, Upton on Severn
George Hudson, Wick, Pershore	Edwin Haywood, Salford
Henry Hudson, Wick, Pershore	H John Walker, Burton Court
W. Clemens, Birlingham	Sir Anthony Lechmere, The Rhydd
F. Woodward, Comberton	H John Racster, Lulsley
J. Winnal, Braces Leigh	H William Rayer, Hillworth
John Lord, Longdon	Charles Bagnall, Lyppard Grange
Willian Coney, Battenhall	H Joseph Smith, Shelsley
J. B. Wagstaff, Worcester	H V. W. Wheeler, Kyrewood House
	W. Woodward. Bredons Norton

Note H = Hereford, otherwise usually Shorthorn

	Sheep	Pigs
L	James Cresswell, Newland	Richard Ashwin, Bretforton
	John Crane Nott, Hallow	Charles Randall, Chadbury
	T. Hawkins, Staunton Court	
	B. Hooke, Norton	
L	J. G. Watkins, Woodfield, Ombersley	
L	F. Woodward, Comberton	
L	Thomas Gale Curtlett, Becere	
	John Dowling, Upper Wick, Worcs	
L	Thomas Harris, Tardebigge	

Note L = Leicester, otherwise usually Ryeland

General Lygon, Mr Onslow, Mr Horneyhold, Edward Holland MP, Lord Lytellton and the Earl of Coventry were early supporters of the Society.

Mr Onslow, seeing that Worcester farmers had to send to Staffordshire for their drills and with a desire to encourage local manufacturers, proposed that a prize be given for the best agricultural implement.

Mr Risling proposed that they extend the horse prizes because 'there was scarce a horse in our fairs fit to sit on'.

A necessary distinction was first marked by Earl Beauchamp and Mr Larkin, who gave prizes for a bull and a cow, which were not Hereford.

In 1845 several of the members were desirous that the Society should try to be more useful and Mr Curtler expressed regret that practical matters did not receive the attention they got from the Evesham Society.

The Society pursued its chequerered career, suffering severe blows in the death of Mr J. C. Nott and Mr Edward Lakin, who succeeded him.

The improvement of the labourer was the original object of the Society and in later years some time was spent discussing the degrading conditions under which Mop or Hiring Fairs were conducted. Sir E. Lechmere suggested that they should be held in some public building and that the clergy and gentry, who took an interest in the morals of the poor, should supervise the occasion. 'The public houses on these occasions should be locked up and not permitted to foster the midnight drunken orgies of the two sexes'.

When the Corn Laws were repealed and ruin appeared to stare farmers in the face, Lord Ward urged them to go for increased quantity, better quality and economy of production. But he did not advise how these objects were to be attained. Sir John Pakington anticipated good results from gold discoveries in California.

If the Society was behind its times in some respects it was not so in respect of steam culture. Mr Curtler offered a 100 guinea prize for a cultivator that could be worked at less expense than a horse machine and Mr Royds produced a steam cultivator of his own invention.

There was much discussion on growing flax, the crossing of Shropshire Down and Cotswold sheep, weights and measures and an attempt was made 'to form a Chemical Society to better understand the science of Agriculture'.

In 1856 and 1857 a two day poultry show was added and held in a large tent. In subsequent years this grew to between twenty and thirty classes and was held in the

Corn Exchange in Angel Street. The prize money during this period ranged between £550 and £650 and dinners were regularly held at the Star Hotel or the Guild Hall Assembly Rooms.

In 1862 attempts were made to amalgamate with the Evesham Society. These negotiations evidently went sour because, in somewhat undiplomatic language, one commentator remarked that the County Society had made a mistake in courting an alliance with a 'lesser luminary'.

The Society was successful in persuading the Royal Show to visit Worcester in 1863 and the Worcestershire show was suspended and all energies devoted to that end.

The Show was held in a field farmed by Mr Foxwell and two fields on the farm of Mr W. Coney of Middle Battenhall, both tenants of Sir Thomas Sebright. The land required for ploughing, mowing and reaping trials was at Nunnery Wood Farm, Middle Battenhall, by the side of the Spetchley road.

The Society offered three prizes for essays:
1. On Hop cultivation, drying and preparing for market, adapted to the counties of Hereford and Worcester.
   Winner J. P. Smith, Lower Wick, Worcester.
2. On orchard management.
3. On Cider management, adapted to the counties of Worcester, Gloucester and Hereford.
   Both won by Clement Cadle, Ballingham Hall, Gloucester.

The City, in addition to a monetary contribution of £2,500, did its best to ensure a successful show. From Shrub Hill station and through the principal streets, shops and houses were gaily decorated.

Conveyance of all kinds crowded the streets on Monday, the opening day. What with Birmingham hansoms, Manchester omnibuses, Worcester, Gloucester, Cheltenham and Malvern flies, private carriages, gigs, tandems, farmers' carts and drays, the horseways were as crowded as the causeways were with foot passengers and curious spectators.

To add to the chaos at around midday, Wombwell's Menagerie entered in procession with band, camels and elephants.

The Alhambra Company also paraded the streets with their band carriage, containing Maori chiefs and one or two natives of 'The Celestial Kingdom'.

No doubt with an eye to the main chance, an agent for 'Simpson's Farinaceous Food' swelled the tide by sending out a monster advertising van.

Prior to the show, twelve sets of steam ploughs puffed through the streets, trailing their ploughs and cultivators.

Of the 75,000 odd who attended, many who came by train had great difficulty in returning home. The railway company was hard put to it coping with the fighting, scrambling crowd on Shrub Hill platforms.

There were no shows in 1865 and 1866 because of rinderpest. In 1867 the show was not as large as had been hoped. It was manifest that the exhibitions were not what they should be and the want of interest in the Society's operations was carefully considered by a committee who, at the next annual meeting, recommended that, instead of confining entries to the county, the show should be thrown open to all England.

The 1868 show was also a disappointment and this was put down to the fact that it was held in the cattle market and that there was no flower show, band or other attraction. President Mr Allsop commented that something had to be done to make the show more attractive and Mr Dowdeswell commented that they must 'either do away with anything like an agricultural society in this county, or they must one and all work for a common purpose to form something more worthy of the great county in which they lived'.

As a result of this and other deliberations it was agreed at a meeting attended by the Mayor, Earl Beauchamp, the Duc d'Aumale, the Earl of Dudley, the Earl of Coventry and numerous other gentlemen of note that the show should extend to three days, that it should be migratory and that a flower show should be added.

Earl Beauchamp accepted the Presidency and the newly organised show for 1869 was held in a large field owned by Mr John Stallard, close to the Henwick Turnpike. At the entrance were a few entertainment booths and it was not until the visitor passed the pay gate that the full extent of the show was visible. The ground occupied by the exhibition comprised twelve acres. The right hand side of the field was devoted to machinery and to the pens of pigs and sheep. Across the top of the field were stalls for horses and in front ground was set apart for exercising them. On the other side of the field were the cattle stalls and fruit, flower, vegetable and poultry tents.

The bands of the Royal South Gloucestershire Militia and the Worcestershire Yeomanry played throughout.

The visitors admitted by payment on the first day numbered 798; 8,000 the second day and 3,000 the third, the first day's numbers augmented by ticket-holders. There were 123 cattle exhibits, fairly evenly divided between Hereford and Shorthorn, 230 sheep being mainly Leicester and Shropshire.

The pigs were more remarkable for their obesity than their character. Mr Allsop of Hindlip Hall entered four hilts, but the best of the lot died on the road.

From contemporary accounts of other shows round the country it is apparent that this tendency for animals to be over-fattened for showing was a problem that the more enlightened breeders were constantly trying to overcome. Time and again there are accounts of cattle and other animals being so fat that they could hardly walk out to be shown and it took many years to reverse this trend.

The horse show attracted considerable attention, the entries were numerous and the prizes, exclusive of the special prizes for jumping, amounted to £230.

In the machinery section the local firms of Humphries of Pershore exhibited threshing machines, J. Larkworthy & Co exhibited ploughs and cultivators and H. Burlingham & Co of Evesham were there too.

'As you passed through the flower tent where the roses lay you found yourself dreaming about some ideal being. If the particular "She" had a complexion anything approaching the colour of the roses, it would be worth all your efforts to get her under the mistletoe in the drawing room next Christmas.'

Notwithstanding an attendance of over 10,000 the show was not a financial success. The committee had to fall back on the guarantee fund. It was rightly considered that a repetition of such generosity should not be relied upon and it was decided that whichever town the show was to visit should raise the sum of £300 towards expenses.

Kidderminster was the first to receive the show under the new conditions and gave it a most hearty welcome, contributing a considerable sum in addition to that required.

In 1871 Malvern likewise gave the show a cordial welcome. A most successful event was spoiled by a high wind on the last day, which blew down the flower tent and took the coverings off the cattle shed, sheep pens, poultry and refreshment tents.

The 1872 show was held at Stourbridge and the local committee raised £736, a larger sum than ever before contributed. The show was held in a 22 acre field at the back of Amblecote Church. A special advantage was the contiguity of the field to the branch railway line and the railway erected a temporary platform, so that trucks containing animals and implements could be unloaded.

The jumping ring was 550 feet long by 120 wide and there is a first mention of a grandstand, which was 200 feet long. There was scarcely a house in the principal streets which did not show flags and bunting. The most effective decoration was a castellated arch across the road by the town clock. The cost of this was defrayed by public subscription.

Cattle entries were down because of foot and mouth disease. Shorthorns occupied the place of honour and first prize bull (there were no championship prizes) was won by Mr Linton of Sheriff Hutton, York. He had previously won at the Bath and West, Essex and Highland shows, as well as winning fifty other awards. Exhibitors, even in those days, took their stock long distances. There was of course an extensive railway system, reaching to the most distant towns and villages.

The sheep and pig classes were well filled and on the implement stands there was everything from threshing to sewing machines. The band of the Grenadier Guards played throughout.

At dinner President Lord Lyttelton commented that a country like England, crowded with inhabitants and yet apparently not by any means arrived at the limit of its resources in any respect, would be guilty of the most suicidal policy if any any time it failed to pay the greatest attention to the development of its own agricultural products, and that a nation's capital was the health and proper use of its land.

He made reference to some recent unrest among agricultural workers and expressed the wish that, as the lot of the worker was now satisfactory, in no respect might it ever deteriorate.

He was talking about a meeting addressed by Joseph Arch at Wellesbourne in Warwickshire. Two thousand farm workers attended and they passed a resolution to form a union. Within three weeks the membership was five thousand and by the end of the year it had reached seventy thousand. Landowners and farmers were amazed at the strength of the movement, which demanded a rise in wages to sixteen shillings a week from twelve.

The *Labourers' Union Chronicle* of 1872 mentioned many instances of bullying, eviction from cottages and witholding of Parish Charities. The clergy for the most part stood aloof or opposed the Union, the nonconformity of Arch not being to their liking. The Bishop of Gloucester said he should be ducked in the horsepond, but there were some in the Church who supported him. The Herefordshire Union was encouraged by several of the local clergy, notably Canon Girdlestone and James Fraser, who later became Bishop of Manchester.

The 1873 show was held at Evesham, the 1874 at Dudley and the 1875 at Worcester, where a horse showing class was introduced. At the meeting the chairman of the committee, Mr G. E. Martin, commented that there were agricultural shows in all the neighbouring counties and some form of amalgamation should be considered. He went on to say that the show had been around all the principal towns and he did not think they would be so successful the second or third time round.

There was no show in 1876 because the Royal was held at Birmingham and the Bath and West at Hereford. In 1877 it returned to Kidderminster after considering cancellation because of an outbreak of cattle plague.

The £300 that had to be guaranteed was soon forthcoming as well as an additional £200 together with other encouragement in the shape of local prizes. The dinginess of the lesser streets was hidden by various decorations. Public buildings had received special attention, including the newly-built Town Hall and Municipal Offices.

Value of the prize money had now risen to £20 for the champion buill, with £10 and £5 respectively for first and second. Cows in milk had to be satisfied with an £8 first prize and yearling heifers £4.

The fact that all the pigs could not only stand up, but that they could look the onlooker in the eye was also commented upon. There had been previous occasions when they were so fat that they could do neither.

There was a class for best farm, in which the first prize was £50 won by Mrs Smithin of Wadborough Park; second, £30 won by Mr J. S. Walker of Knightwick Farm, and Highly Commended, Mr G. R. Essex of Leigh Court.

Twenty-two acres on the Warwick Hall Estate, between the railway station and the town of Bromsgrove, were the site of the 1878 show. The town raised £630 and there were nine classes of Shorthorns, eight of Herefords, eighteen sheep, ten pigs, eighteen horses and butter, implement and flower classes.

The 1879 show was held at Malvern and the principal item of attraction in the implement class was 'Howard's Steam Plough', which was reckoned to be capable of ploughing six or seven acres on a fair day.

The Society combined with the Bath and West in 1880, went to Stourbridge in 1881 and Dudley in 1882. The comment here was that the Gloucester Wagon Company, being the erectors of the temporary structures, might with a little more imagination and at no additional expense make the showyard considerably more attractive.

The cattle, horse and pig classes were well filled but there was a shortage of sheep. Because of rain prior to the show the ground was in 'an awful pickle' and a good deal of the steam tackle could not be got on site in time for the opening.

The horticultural exhibits were a great attraction, the stove and greenhouse plants deserving the highest praise.

The Aylesbury Dairy Company's tabernacle was a department in the show, arrangements having to be made to deal with 30 gallons of milk daily.

The 1883 show at Worcester was something of a disaster; an outbreak of foot and mouth disease occurred close to the showground. After considering cancellation they decided to carry on with additional horse classes. The Aylesbury Dairy Company were induced to bring down additional machinery and a Fire Brigade competition was held.

There was a violent storm when they sat down to dinner and the tent leaked to such an extent that many had to put up their umbrellas.

Foot and mouth again affected the show at Pershore and following year. That and the weather caused a loss of £150. Negotiations for amalgamating with one or two of the adjoining counties dragged on without making much progress. Obviously the organisers were in a dilemma as to whether to carry on or call it a day. However, they were encouraged by a cordial invitation from Redditch and the promise of £250 in the guarantee fund to hold a show in 1885.

Entries in all sections were short. One innovation was an award for portable silos, which was won by the Ensilage Press Company of Leicester.

Instead of a dinner, which had been customary, they had a meeting of members because 'latterly they had found that instead of people being anxious to come they had had to canvas to fill the empty tables'.

The railway company 'except at the risk of the Society', refused to run special trains from Evesham and Alcester. One way or another during the previous three or four years the Society had been fighting an uphill battle. Redditch was the last show.

It could be that people preferred the atmosphere of their own local show, because during this time there were a number of small societies active in the county, notably Bromsgrove Farmers Club, Chaddesley Corbett Farmers Club, Clifton on Teme Agricultural Society, Knightwick and District Farmers Club, Madresfield Farmers Club, Ombersley Agricultural Society, Tenbury Agricultural Society and Great Witley and District Farmers Club.

To some extent the methods of the Society may have been taken up by the Worcester Chamber of Agriculture, who in 1894 moved the idea of amalgamation with Hereford.

Finally a 'Declaration of Trust' dated 14 May 1906 states that 'The Worcestershire Agricultural Society ceased to exist in 1886 but was revived in 1894 and amalgamated with the Hereford Society to form the Hereford and Worcester Agricultural Society and the sum of £336 18s 7d be transferred by the Trustees of the old Worcester Society for the benefit of the Hereford and Worcester Agricultural Society'.

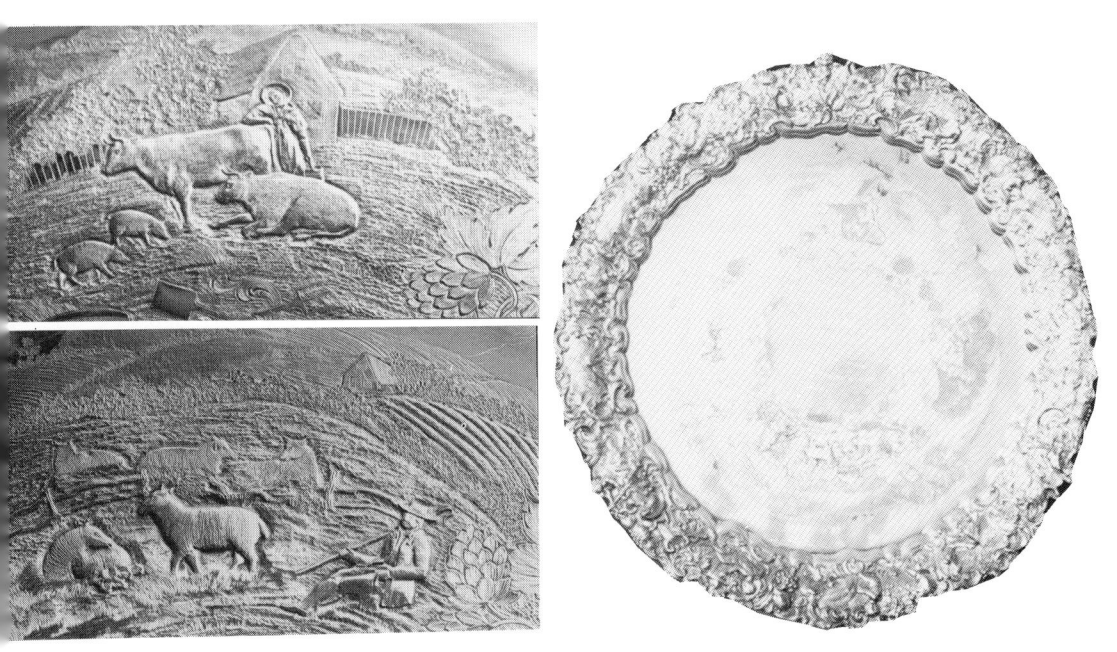

LEFT: Cattle and sheep dominate the engraved detail in the 'Nott' salver, RIGHT: now in Worcester County Hall.

ABOVE: Horse-drawn plough from the Nott salver. CENTRE: A Cotswold ploughing match, c1890. BELOW: Mechanisation: Fowler's steam plough.

Merger In Chamber —
Worcester joins Hereford

The Hereford Society visited various towns by invitation, but the show was growing and it was becoming difficult to find a sufficient number of towns in the county, or not too far away, that could accommodate it.

Because of this Worcester city was asked to have the show in 1894. The city was not too keen, but the Worcester Chamber of Agriculture said it would support the move, if the Hereford Society would agree to amalgamation.

The Central and Associated Chamber of Agriculture was formed in 1866 to represent the political interests of farming and there was a branch in most counties. Because the suggestion came from the Worcester Chamber, it is clear that there was no Worcestershire Agricultural Society in existence at that time.

A committee meeting was held at the Green Dragon Hotel, Hereford on 18 July 1894, to consider amalgamation and to meet a deputation from the Worcester Chamber.

A Mr Haywood, the Vice President of the Hereford Society, was in the chair and in his opening remarks he commented 'It seems to me there is a growing body of persons in Worcestershire, who are very desirous of having us, which has been shown mainly through the Chamber of Agriculture. But we have got some — I won't use the word deadly — opponents, some of whom are rather bitter against us, who it seems are not going to approach us unless we have an idea of amalgamation'.

In the debate which followed it emerged that the Society could not go to Worcester, or anywhere else in the county, unless an amalgamation was agreed.

Rev Davenport thought that the Hereford Society had grown too big for the county, its towns being too small to accommodate them, with the exception of Leominster. In any other town they lost money. It was noted that when it had been proposed to hold meetings in Worcester and Gloucester there had been jealous feeling. They had no society of their own and there had been a 'dog in the mangerish' feeling. 'We have not got a society of our own and we don't want you'.

There is no doubt that in those days they believed in plain speaking!

The feeling was that Worcestershire, not having a viable society, this was a good time to amalgamate. In the discussion it was regretted that Gloucestershire had a society up and running, otherwise it would have been an opportune time to bring them in as well.

Mr Britten thought it would be a waste of time to approach Gloucestershire, who had snubbed them a great deal more than Worcester, when they were last approached.

The Worcester delegation was than asked into the room and they informed the committee that it was the unanimous wish of the Worcester Chamber and many farmers in the county that amalgamation take place.

This being the feeling of everyone the resolution was put to the general meeting of members on 29 August 1894 and carried unanimously.

At a subsequent meeting it was agreed that the joint society be called the Hereford and Worcester Agricultural Society. Four stewards were to be appointed from Worcester and ten further members on the Council, making fourteen in all, the Lord Lieutenant and the Mayor of Worcester to be additional *ex officio*.

Earl Beauchamp was elected first President of the new Society, with two Vice Presidents, one from each county. It was also agreed that £1,000 be put into a special fund. This was to be entirely under the control of the Hereford members, the income to be used for the promotion of Hereford cattle.

The first joint show was held on Worcester race course on 18, 19 and 20 June. The members of Council were not imbued with the same sense of urgency pertaining today, because the prize list committee did not meet to select judges until 10 April of the same year as the show.

In November 1885 a letter was read from the Central Chamber of Agriculture inviting the Society to join in a joint approach to the Board of Agriculture to try to enforce the slaughter of imported animals before embarkation, as a disease precaution.

A lot of time was taken up at council and other meetings hearing complaints, invariably from horse exhibitors, that they had been disqualified or not given a prize. They were often represented by solicitors. The reasons for complaint were that they had won a prize at some other show, therefore should have won at this. The cattle exhibitors and others seemed more inclined to accept judges' decisions as final.

At the first meeting after the amalgamation it was reported that the Society's membership had risen from 455 to 722 and there was a credit balance of £1,134 19s 3d.

The centenary show was held at Hereford on 14, 15 and 16 June. It was suggested that a dog and poultry section be added but, after much discussion, it was decided to proceeed with poultry but not with dogs.

At a meeting on 19 February 1898 Mr T. H. Edwards was appointed as assistant to the Secretary, his father Mr Alfred Edwards, thus maintaining a long and notable contribution to the affairs of the Society until Mr Glynne Hastings was appointed in 1946.

A proposition was put forward to the Board of Agriculture at around this time: 'The Hereford and Worcester Agricultural Society having heard of a proposal by the London County Council to take steps with a view to the abolition of private slaughterhouses in London and the substitution of public slaughterhouses, is of the opinion that the adoption of this proposal would be most injurious to the interests of British Agriculture by limiting the demand for home fed stock'.

At this distance the logic of this proposal is a little difficult to follow, but it does show that the Society spent time debating political issues.

At a meeting on 18 June 1902 Council received with regret the resignation through ill-health of Mr Alfred Edwards, Secretary for 21 years. It was unanimously resolved that his son Mr T. H. Edwards be appointed in his place.

Speaking at the Presidential banquet at Stourbridge in 1899, Lord Cobham commented on the successful amalgamation of the two societies; Worcester had been on the decline for some years, there were times when they had not been able to hold a show at all. He put this down to the fact that a handful of people had done most of the organising. In the fullness of time they had come to retirement and no-one had been willing to take their place.

Amalgamation meant that there would be more choice of venues for the show and, reading between the lines, it would seem that some towns did not relish being visited too often. To get over this difficulty the show travelled as far afield as Monmouth, Brecon and Radnorshire.

At the turn of the century the showground covered some twenty-five acres and at Stourbridge for the first time there were two avenues of machinery. The grandstand was three hundred feet long and for ring events they had jumping, a pairs turnout, single turnout, a pony turnout and a polo pony bending race. Polo matches were a regular feature of the ring events.

During the ten years either side of 1900 average cattle entries were 158, consisting of Herefords, Shorthorns and the occasional Jersey; sheep 82, being mainly Shropshire and Ryelands; 12 pigs, mostly Berkshires, and 180 horses, fairly evenly divided between heavy horses and hunters.

There were averages of 42 entries in the cider and perry classes, eight in the wool and 28 in the butter and cheese.

In the event that a judge could not make up his mind when the exhibits were of apparently equal merit, it was usual for lots to be drawn for first place.

After a visit to the Bath and West Show by various members of the Council, forestry classes were introduced in 1908.

An 1890 catalogue of the Hereford Society mentions W. Wargent of Canon Frome. By 1903 they had become W. Wargent & Son of Canon Frome and the *Hereford Times* spoke of 'This enterprising firm, their stand attracting the attention of prominent agriculturists. In addition to mowers, binders, horse rakes, turners and haymakers by such leading firms as Harrison McGregor, Bamfords, Massey Harris, Nicholson and Ransomes, Messrs Wargent exhibit a hop scuffle of their own making. It is a handy and compact implement manufactured at the ridiculously low price of £8 and is regarded with much favour by hop growers generally. At the front of the stand there were two Blackstone oil engines running. They are regarded as the farmer's engine and can be used for chaff cutting, corn grinding and all general work'.

Still exhibiting today, Messrs Wargent are far and away the longest serving exhibitors.

A spectacular display always helps make the show 'go', but the engagement of some of them was not without hazards.

In 1910, when the show was held on Worcester racecourse, the star attraction was the visit of a Bleriot-type monoplane. The crowd was so excited that, when the aeroplane tried to take off, they became uncontrollable. Despite repeated pleas from one 'Diabolo', the mechanic, they kept pushing forward on to the runway. The plane was likewise uncontrollable and veered off course into the crowd, finishing upside down in the entrance to the flower tent. The pilot, Captain Clayton, was thrown clear and suffered no injury, but several of the spectators were injured and a Mrs Eileen Pitts was killed.

The Council spent a considerable amount of time over a long period discussing this tragedy. In all it cost the society around £1,000 in compensation, the husband of the unfortunate lady who was killed being awarded £150.

Between 1900 and 1910 hay prices for the show were quoted at up to £5 per ton, straw £3 10s and vetches, for green feed, £1 10s.

When selecting sites for shows, distance from the railway station was an important factor, as the main form of transport for freight, animals and people.

In 1908 classes for the best farms were introduced:
1. Farms of 200 acres and upwards 1st £60, 2nd £30, 3rd £15.
2. Farms 150-200 acres of which not less than 20% under Hops and Fruit 1st £40, 2nd £30, 3rd £10.
3. Farms of 50-100 acres, mixed arable and pasture 1st £40, 2nd £30, 3rd £10.
4. Farms 10-15 acres chiefly devoted to fruit growing and market gardening 1st £20, 2nd £10, 3rd £5.

In 1909 forestry classes were introduced.

Also considered were team jumping and riding competitions in the hope that Yeomanry Regiments would be interested.

The Council meeting of 4 October 1913 received a deputation from the Bath and West of England Agricultural Society, who wished to visit Worcester in 1915. It was agreed to suspend the Hereford and Worcester show for that year and to co-operate with the Bath and West. Worcester city agreed to find a forty acre site and £800 towards running the show. The Hereford and Worcester Society agreed to find £100 which was eventually raised to £300, for classes of their own choosing, provided they came within the general objectives of the Bath and West.

The 1914-18 war intervened and the Bath and West was held in a much reduced form. The 1916 show was to have been at Evesham but, because of the war, the invitation was withdrawn and it was held in a much reduced form at Hereford. It was felt that holding the show on a purely agricultural basis, without the peripheral stands and events, would be beneficial to the drive for increased food production.

The Council was invited to nominate one member to serve on the War Agricultural Committee and this was eventually increased to three.

They then met on 19 April 1916 to consider a request from the Ministry of Munitions not to hold an exhibition of implement and machinery, under war-time regulations. It was decided to cancel the show and instead hold a one-day event in Hereford market and the adjoining meadow.

Despite the fact that a number of subscriptions had been lost, because many members were on active servcie and L. H. Woodhouse had to be compensated in the sum of £110, for expenses already incurred, there was nevertheless a profit of £66 12s 4d. This result has to be viewed in the light of the fall in the value of the Society's investments, which at that time were put at £337 15s 11d. This put the total value of all the Society's assets at £3,167 4s 1d.

After strong representations had been made to the Board of Agriculture and the Ministry of Munitions, who refused to give permission, the 1917 show was cancelled. Council was given to understand that no show would be allowed so long as the war lasted.

Instead the Society introduced classes for the best managed dairy herd, to be judged on the farm: Class 1 for herds of 16 or above and Class 2 for herds of 8 to 15 cows.

At a general meeting on 20 February 1918 two resolutions were passed, first 'that in the opinion of this meeting the system of selling cattle and sheep by live weight is best. The proposed new system of selling by dead weight in Government slaughterhouses is not likely to operate in the general interests of farmers'.

The second resolution was to ask the War Agricultural Committee to discontinue the ploughing up of pasture land in Hereford, as much of the land was only suitable for rearing stock. The continuance of ploughing up would operate unfavourably to the rearing of pedigree Hereford cattle.

Although by 1919 the war had come to an end, costs had in many cases gone up by 400%. It was doubtful if implement manufacturers, who had been making munitions, would be able to exhibit. It was therefore decided to hold a one-day show in Hereford market on 12 June 1919.

One letter was read from Mr T. J. Salway offering the Salway Cup, which had been awarded in 1807 to a bull called Andrew belonging to his grandfather.

Another letter read at a meeting on 21 February 1920 brings a wry smile to present day faces. It was from the postmaster, apologising for some of the meeting notices being delivered late.

At a Council meeting on 25 May 1921 there was some discussion on postponing the show, because of the coal strike. They decided to carry on and were informed that entries of livestock were 573. 6,400 feet of shedding had been erected, subscriptions received amounted to £605, and entry fees for stock £345 and for implements, £1,028.

Scene at Hereford and Worcester Show, 1908.

To all whom these Presents shall come to or concern **William Stallard** of the City of Worcester Land Agent **Herbert Wilson Buck** of the same City Land Agent **Sir John Richard Geers Cotterell** of Garnons in the County of Hereford Baronet and **Henry James Bailey** of Rowden Abbey Bromyard in the said County of Hereford Esquire **send Greeting Whereas** a sum of Two hundred and ?? ???? pounds and five shillings belonging to the old Worcestershire Agricultural Society which ceased to exist in or about the year One thousand eight hundred and Eighty six but was revived and became amalgamated with the Herefordshire Agricultural Society in or about the year One thousand Eight hundred ???? and to be applied for the purposes of the said Society in such manner as the Council of the said Society shall from time to time direct **In Witness** whereof the said William Stallard Herbert Wilson Buck Sir John Richard Geers Cotterell and Henry James Bailey have hereunto set their hands and seals the fourteenth day of May One thousand Nine hundred and six

Signed Sealed and Delivered by the above named William Stallard and Herbert Wilson Buck in the presence of

E. ????
Surveyor ????

Signed Sealed and Delivered by the above named Sir John Richard Geers Cotterell in the presence of

Signed Sealed and Delivered by the above named Henry James Bailey in the presence of

Hereford & Worcester deed of amalgamation.

History of the Society's Investments.

1894 The amount standing at the credit of the old Herefordshire Agricultural Society, previous to amalgamation was £1190 19 11

1895 At the amalgamation in 1895 it was resolved as follows:—

No 1 %. That the sum of £965 12 8 India 3½% Stock then held by The Herefordshire Agricultural Society, being part of the accumulated reserve fund be retained and the stock transferred into the names of three gentlemen, members of the Herefordshire & Worcestershire Agricultural Society, who are resident in the county of Hereford, and that the annual Income thereof should be applied in giving Special Prizes for Hereford Cattle or in some other way for the advancement of the breed of Hereford Cattle, such invested fund to be entirely under the control of the Herefordshire Members of the society, and subject thereto that the funds of the Herefordshire Agricultural Society be available for the general purposes of the Herefordshire & Worcestershire Agricultural Society.

OFFICERS.

President:
THE RIGHT. HON. LORD HINDLIP.

Vice-Presidents:
W H. DAVIES, Esq., and H. W. BUCK, Esq.

Council:

THE PRESIDENT.

THE VICE-PRESIDENTS.

THE STEWARDS.

THE LORD LIEUTENANT OF HEREFORDSHIRE.
(Sir John G. Cotterell, Bart.).

THE LORD LIEUTENANT OF WORCESTERSHIRE.
(The Right Hon. The Earl of Coventry).

THE MAYOR OF HEREFORD.

THE MAYOR OF WORCESTER.

ALLSEBROOK, A., B.Sc. (Edin.), F.S.I., Madresfield Grange, Malvern.
AMPHLETT, Thomas, Brockhampton, Stourpourt.
BARNEBY, W. T., Saltmarshe Castle, Bromyard.
BEAUCHAMP, Right Hon. Earl, Madresfield Court, Malvern.
BEDDOE, H. C., Castle Street, Hereford.
BRITTEN, Rear-Admiral, R. F., Kenswick, Worcester.
BAILEY, Henry J., Rowden Abbey, Bromyard.
BROOKS, J. B., Finstall Park, Bromsgrove.
COTTON, E. B., Sunnymead, Bromsgrove.
*DAVIES, W. H., Claston, Hereford.
DAVIES, J., Hope House, Martley, Worcester.
*EDWARDS, C. T., Byford Court, Hereford.
EVANS, Francis, The Weston, Bredwardine, Hereford.
*GERRARD, G. Overton, Hindlip, Worcester.
GREEN, G. H., Wigmore Grange, Leintwardine.
*HICKMAN, W. H., Broome, Stourbridge.
HEYGATE, Capt. E. L. A., Buckland, Leominster.

Council (continued).

*HUGHES, A. E., Wintercott, Leominster.
HYSLOP, J. K., Chipps House, Ivington, Leominster.
*MOORE, Lawton L., Brampton Brian.
MILLYARD, J. W., Little Bridge, Bromyard.
NOTT, Charles, Fairfield, Kingsland, R.S.O.
*PULLEY, Charles T., Lower Eaton, Hereford.
PITCHER, G. W. Y., Little Comberton, Pershore.
*RANKIN, Sir James, Bart., Bryngwyn, Hereford.
*ROBINSON, Stephen, Lynhales, Kington.
ROWLANDS, Rolla, Evesbatch Court, Bishops Frome, Bromyard.
SMITH, James, Monkton, Hereford.
TURNER, A. P., The Leen, Pembridge.
TAYLOR, H. W., Showle Court, Ledbury.
*VERNON, Sir Harry, Bart., Hanbury Lodge, Droitwich.
WARD, R. Bruce, Westwood Park, Droitwich.

* Retire by rotation, but are eligible for re-election.

Stewards of Stock:

BUCK, H. W., Pierpoint Street, Worcester.
CADDICK, E. W., Caradoc, Ross.
CONEY, Frederick, Besford, Worcester.
JONES, J., Newton, Bradley Green House, Redditch.
YEOMANS, J. H., Stone House, Withington, Hereford.
POWELL, John, Lower Wick, Worcester.
SMITH, J. W., Thinghill, Hereford.
WALKER, T. Lawson, Knightwick, Worcester.
WILLIAMS, Christopher, Holmer, Hereford.

Stewards of Finance:

JOSELAND, Charles, Bank Buildings, Kidderminster.
STALLARD, Col. W., St. John's House, Worcester.
HAYWOOD, W. M., Westfield House, Hereford.

Director of the Yard:

EDWARDS, Dearman, Brinsop Court, Hereford.

Bankers:

THE NATIONAL PROVINCIAL BANK OF ENGLAND.

ABOVE: Hereford's statement of assets at amalgamation. BELOW: List of Officers and Council of Hereford and Worcester Society.

ABOVE: Opening ceremony of Hereford and Worcester Show at Hereford in 1907. The show was opened by the President, Lord Chesterfield. BELOW: Hereford and Worcester show at Malvern in 1914. INSET: Declaration of Trust costs to transfer Worcester's assets to the Hereford and Worcester Society.

ABOVE: Dignitaries at Hereford and Worcester Show, held at Malvern in 1905. BELOW: The Show at Bromsgove, 1906.

MEMBERS OF THE ASSOCIATION.

* BEAUFORT, HIS GRACE THE DUKE OF, Patron.
* SEGRAVE, THE RIGHT HON. LORD, President.

* Bubb, Anthony, *Bentham*
* Barnes, Edward, *Tirley*
Beach, John, jun., *Redmarley*
Baker, T. J. Ll., *Hardwick Court*
Baker, Thomas Barwick L., *Ditto*
Barnard, Joseph, *Apperley*
Berkeley, Grantley F., Esq., M.P.
* Brown, Samuel, *Kingstanley*
Butcher, William, *Standish*
Caruthers, E. P., *Brown's Hill*
Cole, Jas. Stratton, *Gloucester*
Chadborn, John, *Ditto*
* Codrington, C. W., *Dodington Park*
Cambridge, C. O., *Whitminster*
* Clifford, H. C., *Frampton*
* Canning, Robert, *Hartpury*
* Colchester, M., *The Grange*
Crump, Thomas, *Ashelworth*
* Crump, John, *Hasfield*
* Dancocks, H. H., *Dymock*
Dowling, J., *Gloucester*
Eycott, Frederick, *Stonehouse*
Evans, W. S., *Prestbury*
Ellis, W. V., *Minsterworth*
Ellis, W. J., *Berkeley*
* Goodrich, W., *Matson House*
Guise, Sir W., *Rendcombe Park*
Hayward, D. S., *Frocester*
* Hayward, J. Curtis, *Quedgley*
* Hawkins, T., *Staunton Court*
* Hawkins, Wm., *Hawthorns*
* Hicks, Sir Wm., *Witcomb*
Hope, H. T. Esq., M.P.

* Hyett, W. H., *Painswick*
Hulls, Thomas, *Corse*
Hopton, James, *Wotton*
Helme, M., *Pagan Hill*
Holtham, W., *Gloucester*
Hill, Rev. Thomas, *Redmarley*
Hawkins, Jeremiah, *Tirley*
Ireland, George, *Eldersfield*
* Jones, Samuel, *Gloucester*
* Jones, William, *Sheep House*
* Jones, Rev. Edward, *Hay Hill*
* Kemp, J., Secretary, *Gloucester*
* Lawrence, W. L., *Sandywell Park*
Leversage, Peter, *Stroud*
* Lewis, W., *Lypiatt*
Lane, John, *Deerhurst Walton*
* Martin, Richard, *Whitminster*
* Marling, N. S., *Stroud*
Mathews, P., *Dunsbourne*
Marklove, Charles, *Berkeley*
Marling, T. P., *Stroud*
Morris, W., *Maisemore*
* Niblett, Daniel J., *Haresfield*
Nind, Charles, *Hailes*
New, John, *Northway*
Phillips, Samuel, *Coldthrop*
* Priday, William, *Longford*
* Partridge, J. W., *Bowbridge*
Pike, Aaron, *Mitton, near Tewkesbury*
Proctor, Michael, Esq., *Tewyning*
* Ricardo, D., *Gatcomb Park*
Richards, —, *Dumbleton*
Rayer, John, *Twyning*

* Smith, B. Charles, *Whaddon*
* Stanton, W. H., *Stroud*
* Stallard, Joseph, *Redmarley*
* Smith, Thomas, *Whaddon*
Staunton, W., *Thrupp*
Smith, P.
* St. Clair, D. L., *Staverton*
Smith, Thomas, *Gloucester*
* Tolley, Charles, *Ditto*
* Tolley, Thomas, *Twyning*
* Talbot, George, *Temple Guiting*
Toohy, Wm., *Hartpury*
* Trye, H. N., *Leckhampton*
Tasker, C. J., *Gloucester*
Turner, Thomas, *Ditto*
* Viner, Jos. Ellis, *Badgworth*
* Walker, D. M., *Gloucester*
Walters, J.W., Treasurer, *Barnwood*
Wintle, James, *Saint Bridge*
* Whitcombe, H. S., *Longford*
* Whitcombe, J. A., *Gloucester*
* Whitcombe, Capt. T. Douglas, *Gloucester*
* White, John, Bagnard *Sand Linne*
Whithorn, Henry, *Tredington*
* Walker, John, *New Hall, Chaceley*
* Watts, Joseph, *Stroud*
Weedon, E., *Gloucester*
Wyatt, H., *Stroud*
Wilton, Robert, *Gloucester*
Wilton, W., *Ditto*
Washbourn, W., *Ditto*
Webb, Edward, *Adwell*

Those marked thus * are Members of the Committee.

JOHN KEMP, Secretary.

That the following be the order of proceedings at all General Meetings:

First.—The books of rules and orders of minutes and correspondence shall be laid on the table before the Chairman.

Second.—None but members shall be admitted to the meetings of the Association, unless introduced by a member.

Third.—When any member speaks, he shall rise and address himself to the Chair; and if two members speak together, the Chairman shall call them to order and decide which shall speak first.

Fourth.—When any matter is in debate, if a member shall speak to new business, the Chairman shall call him to order.

Fifth.—No debate shall be entered into or question put on any motion, unless that motion be seconded.

Sixth.—No motion that has been rejected shall be made again the same day.

N.B. As the proper and regular dispatch of business at the General Meetings will very much depend on the diligence and attention of the several Committees, it is respectfully requested, that the Gentlemen appointed thereon will give as general attendance as possible, both at the sittings of the Committees, when summoned, and at the Meetings of the Association; and that they will meet as nearly as they can at the hours appointed.

Membership list, order of proceedings, and an exhortation to members
of the Gloucestershire Agricultural Society in 1833.

A Respectable Assemblage —
The Gloucestershire Agricultural Society

The first mention of a Gloucestershire Society appeared in the *Gloucester Journal* of 2 May 1829. There was a comment that the Cirencester and Gloucestershire Agricultural Association had published a list of premiums for a show. The writer further commented that the formation of such a society was long overdue.

The show was advertised to be held at Cirencester on 1 December 1829. The *Journal* reported that 'a more numerous or respectable assemblage of gentlemen and agriculturists had seldom been witnessed.' The judges were Mr Wells of (H)Ampnett, Mr Bennet of Caddleworth, Mr Edmonds of Kelmscott, Mr Hewer of Aston and Mr Hook of Wotton under Edge. Having inspected the stock they left the cattle yard to deliver to the committee their award of premiums. The yard was then opened for admission and in a few minutes crowded to excess.

'At four o'clock the company repaired to the Ram Inn where one hundred and thirty sat down to an excellent dinner. The chair being offered to Earl Bathurst, who declined it, it was taken by George Talbot Esq, who communicated to the company the deep regret of the President Lord Sherborne at having to be absent in London that day.

'The Hon Henry Moreton rose and said that instead of limiting the society to the County of Gloucester and twenty miles around, he would propose the admission of stock from any part of the United Kingdom. This proposal was seconded by Joseph Cripps and carried. A number of fresh subscribers immediately entered their names.'

Mr Cripps then presented the prizes to the successful winners, who were:

Mr Young of Kemble, best fat steer	Hereford	£5
Mr Keene of Braydon House, best fat cow	Hereford	£5
Rev W. George, Cherrington, best fat heifer	Hereford	£5
Lord Sherborne, Dairy Cow	Durham and Hereford	£5
Mr Young, Kemble, two-year-old bull	Hereford	£5
Mr Ruck, Down Ampney, aged bull	Hereford	£5
Mr Large, Broadwell, Shearhog ram	Improved Cotswold	£5
Mr Barton, Coln Rogers, three fat wethers	Cotswold	£5
Mr Large, Broadwell, three fat ewes	Cotswold	£5
Mr Slatter, Stratton, ten store ewes	Cotswold	£5
Mr Hall, Coates, ten fat shearlings	Cotswold	£5
Mr C. Stephens, Port Farm, ten theaves	Cotswold	£5
Mr Powell, Yanworth, Cart Stallion		£5
Mr Ruck, Down Ampney, two-year-old cart filly		£5
Mr Betterton, Ashton Fields, Boar Hog		£5
Lord Sherborne, Sow Pig		£5

Although according to this report the Society got off to an apparently good start, the following and subsequent years it is only referred to as the Cirencester Association. Most likely the general depression in agriculture at the time stifled the birth of a county society. Times were certainly hard because, in the same column that the show was reported, it is mentioned that a John Jertin Esq of North Nibley returned ten per cent of rent to his tenants and Sir R. C. Hoare of Stourhead returned fifteen per cent of rents to his agricultural tenants.

The next we hear is that a notice was printed at *The Chronicle, Gloucester*, stating that: 'At a very numerous meeting of Gentlemen, Landowners, Farmers and others, held at the Spread Eage Inn, in the city of Gloucester on Saturday 28th December 1933, rules were drawn up for a "Gloucestershire Association for the encouragement of Agriculture, Arts, Manufactures and Commerce".'

Mr T. J. Lloyd Baker was in the chair, supported by such eminent names as Grantley Berkely MP, C. W. Codrington, H. C. Clifford, Sir William Guise, Sir William Hicks, H.T. Hope MP, T. P. Marling and approximately another hundred gentlemen of note. Mr John Kemp was elected secretary.

The following resolutions were agreed:

'1. That a society be formed, to be entitled The Gloucestershire Association for the encouragement of Agriculture, Arts, Manufactures and Commerce.

'2. That the objects of this Association be — The promotion and encouragement of Agricultural knowledge and practice — The improvement of stock — the bettering of the condition of the agricultural labourer — and the support generally of the arts, manufactures and commerce; and that the Landowners, Farmers, Manufacturers and inhabitants of Gloucestershire and the neighbouring counties, be invited to lend their aid and co-operation in effecting these purposes.

'3. That His Grace the Duke of Beaufort, the Lord Lieutenant of the County, be requested to become the Patron of this Association.

'4. That the Right Honourable Lord Segrave be requested to accept the office of President.

'5. That the nobility of the county, the members of Parliament for the County and City and Boroughs within the county, be requested to support the objects of the Association.

'6. That James Woodbridge Walters be appointed Treasurer; and that Mr John Kemp be requested to undertake the office of secretary.

'7. That books shall be immediately opened for receiving signatures and subscriptions and that the same do lie at the National Provincial Bank, at the offices of the Gloucester Journal and Gloucester Chronicle, at the Spread Eagle, Northgate St; the Fleece, Westgate St, Gloucester; at the Lamb Inn, Cheltenham and at the different banks in the County.'

Subscriptions were fixed at not less than a pound, a benefaction of not less than twelve pounds for life membership, two pounds for a vice president and twenty four pounds entitled a person to be a vice president for life.

At a meeting on 22 February 1834 a schedule of classes was drawn up: seven classes for dairy stock, three for fat cattle, six for sheep, three for pigs, two for horses, two for cheese, two for the best type of plough, one for an implement most useful in any type of agriculture, two for bringing up the biggest family without Parish assistance, one to the exhibitor of the best piece of cloth made entirely from British wool and finally four awards for ploughing.

It is uncertain where this show was held, but it was most likely not too far removed from the Spread Eagle Hotel, which was used for the early meetings.

The ploughing match was held on 14 October 1834, but it must have been something of a flop for it is recorded that the judges gave no prizes, because of 'No merit due'. However, they gave a sovereign to Joseph Parsons, manager of John Lawrence's plough of Churchdown.

Moving on to 15 March 1834, the matters dealt with were entirely of a political nature. They are reproduced in full because they give an interesting insight to the matters which gave cause for concern at the time.

'Resolved: That it is the opinion of this Association that the use of flat soled wheels for wagons, as recommended by the Surveyor, was proved from experiments made this day, to be less injurious to the Turnpike roads and equally applicable to farming purposes, and that the same would be readily assented to by the country at large, provided inducement was held out by enactments for reduction of tolls, with any compulsory provision, which this association consider would be oppressive to the farmer.

'Resolved: That the profits of Agriculture have diminished to so fearful an extent — that the British Farmer will be ruined — that your petitioners are convinced that the distress arises from several causes.

First from the continual agitation of the question of corn laws and that the attempts that are made in various quarters to withdraw even the slightest encouragement to British Agriculture, which these laws afford.

Second From the pressure of the poor laws, aggravated by the influx of produce of all sorts from Ireland, where that pressure is unknown, and from the number of Irish Labourers, who are driven into this country to seek for that subsistence, for which no provision is made for them at home.

Thirdly By the operation of the tax on Malt, by which the cultivation of one of the most important crops is discouraged, Labourers of all kinds are compelled to pay more than double for an article most conducive to the health and comfort of themselves and their families.

'Your petitioners therefore humbly pray that the present Corn Laws continue unaltered, that the burden of the requisite provision for the poor be more equally borne by the different classes of Society and that they be extended to Ireland.

'That the existing Duties on Malt be repealed. Resolved: That this petition be presented to the House of Commons by Sir B. W. Guise Bart.'

At a meeting on 25 November 1834 there was a discussion as to whether a three-and-a-half-year-old heifer, that had never had a calf, should be eligible for a prize as a fat cow. The decision was 'no', but it was recommended that in future animals under four years should be regarded as heifers and over four as oxen.

At the same meeting it was agreed to give an award for the greatest number of children reared without recourse to Parish relief. From this distance what appers to be a somewhat anomolous decision was made to award £2 to James Brown of Haresfield with seven children and £1 to William Evans of Hartpury for eight children.

On 18 July 1835, consideration was given to holding the show in the New Market. This would be what most people now know as the Old Market, which was just off Kings Square and is now occupied by a multi-storey car park.

After some hesitation the Corporation agreed, on the understanding that any accommodation erected for the cattle should be taken down after the show and the site made good.

Ploughing matches were held annually and to our eyes would have been a pleasing sight. On 6 October 1835, first prize went to D. J. Niblett, using a Ransome, Suffolk-type plough, drawn by four horses, two and two abreast. D. J. Hayward of Frocester Court won 2nd with a pair of horses 'without a driver'. He also won with a Beverstone plough drawn by four oxen.

All drivers were given one shilling each.

Classes at the show remained similar over the years. In 1837 classes were added for the six best turnips, six best mangel wurzels, six best cabbages and best one hundredweight of cheese.

In the machinery section there was a class for the best machine for spreading lime and a class for 'A machine most useful in any branch of Agriculture'. Prize for this was to be according to merit and the committee further covered themselves by adding the condition 'According to funds available'.

On 17 December 1836 a deputation was received from the Cirencester Agricultural Society with a view to amalgamation. On second thoughts the Cirencester Society felt they were strong enough to carry on alone, so the negotiations fell through. The occasion is interesting because a Mr D. Trinder was a member of the Cirencester delegation. In later years Mr Edward Trinder was Secretary of the Gloucestershire Society, and also agent to the Bathurst estate.

On 29 December 1838 the Duke of Beaufort offered a prize of £20 — for the best half-bred colt or filly suitable for hunting.

At the same meeting the Earl Ducie offered £20 — for the tenant who had most efficiently drained the greatest quantity of land without help from his landlord. It was not until 29 October 1842 that the judges were able to report that five competitors had entered and they gave the first prize to Harry Parker of Fairford. The method of drainage generally adopted was a stone or tile-built duct with the trench infilled with thick hawthorn.

The competition gave rise to a lot of speculation as to the depth the drains should be put, some favouring drains near the surface and others a depth anything up to six feet. The 'shallow drainers', seem to have no more understood the merits of 'deep drains', than do present-day farmers who have come across these drains from time to time.

Lord Moreton offered prizes for poultry: pair of dressed chickens, pair of dressed turkeys, pair of dressed ducks. These awards were confined to wives or daughters on holdings under fifty acres.

On 13 March 1841 a committee was appointed to meet the Mayor of Gloucester to get the Corporation to erect permanent sheds in the market. In October of that year it was reported that the Corporation had erected two sheds in the market 'adjoining Mr Church's garden'. The Society agreed to pay £30 per annum, so long as they used the market.

The Society continued holding shows in the market but support dwindled due to general uncertainty in the industry. By 1847 several of the classes had to be cancelled because of lack of interest. Also the Society was in debt to the Secretary, Mr Jones, to the tune of one hundred pounds.

This position could partly have been due to a family row between Lord Fitzhardinge the President and his brother Grantley Berkely, MP for West Gloucestershire. Grantley Berkeley was a great supporter of the Society but relied on his brother for the greater part of his inc ome. The argument was so bitter that Lord Fitzhardinge ordered his tenants to withdraw all support from his brother, on pain of losing their tenancies. The dispute went beyond the Berkeley family and many dissensions were caused.

By 1850 the cattle classes were rather better filled but there was a dearth of entries in the sheep classes. In all there were thirteen cattle classes, five sheep, two pigs, one horse, poultry and root classes. There were also long service awards.

In proposing the loyal toast at the dinner the chairman, Sir C. W. Codrington, gave it to our 'Protestant' Queen and further commented, 'you will have little difficulty in confounding the machinations of the Pope and maintaining Protestant ascendancy'. This brought forth loud cheers and was a reference to the Roman Catholic Church re-establishing Bishops in England at that time. Feelings ran high and almost amounted to riot situations in many districts.

There was some discussion on the general economic situation and the Chairman commented that Sir Robert Peel should be brought to account for saying that wheat should fetch fifty six shillings a quarter, when in fact it would only make forty.

There then followed some not very enlightened comments on the position of tenants, most of the company being landlords. Mr Edward Holland then rose and in a clear-sighted address commented that the position of landlords versus tenants was in drastic need of overhaul. If a tenant was give a long lease, this would allow him to raise capital on mortgage, which would enable him to farm better to the mutual benefit of landlord and tenant alike.

In 1852 at the annual dinner he advocated groups of farmers forming co-operatives to send their cattle for slaughter. The offal could then be processed and returned to spread on the land. This source of fertiliser of an estimated value of £1,500,000, 'might bring the suppliers of Peruvian Guano to better terms'.

Edward Holland was a member of Parliament and owner of the Dumbleton estate. He supported both the Gloucester and Worcester Societies by giving premiums, which he frequently won himself; he also persistently advocated the amalgamation of the two Societies. His speeches are still worth reading for their clarity of thought. A keen supporter of agricultural education, he took over the management of the Royal Agricultural College at Cirencester at a time of financial crisis. He was President of the Royal Agricultural Society in 1873.

In 1853 the Royal Show visited Gloucester, Bristol in 1878 and Gloucester again in 1909. On all occasions the Gloucester Society put all their efforts into assisting.

The Cirencester Agricultural Society had been in existence for as long as the county Society and both were attracting mediocre support. They decided to amalgamate and the first show of the new Society was held at Cirencester Park under the Presidency of Earl Bathurst. Others present were Lord Emlyn, Hon Ashley Ponsonby, Rev G. C. Talbot, Edward Holland MP, Messrs Hayward, Cripps, Elwes, Warner, Bowley, Slatter, Hulbert, Stratton, Creed, Mann and Newcombe. The classes were well filled, the poultry section especially.

The new Society held shows alternately at Cirencester and Gloucester. In later accounts Cirencester is mentioned as the parent body. Be that as it may, contemporary reports state that there were numerous agricultural societies of little note in the county: Northleach, Stow, Moreton, Bredon, Winchcombe, Berkeley, Stroud, Painswick.

In 1862 the leading county agriculturists merged together to form the 'Enlarged County of Gloucester Society'. In that year the opening of 'The People's Park' in Gloucester presented a favourable opportunity to bring the remodelled Society before the public.

It was decided that in future the new Society would move around the county to the larger towns. The first show was held in Gloucester, and deemed a suitable event to inaugurate the new Park.

The show was held on Wednesday and Thursday during the last week of July 1862 and was reported in the *Gloucester Journal*:

'Never since the great meeting of the Royal Agricultural Society, nine years ago, has Gloucester assumed so festive an appearance. On Wednesday morning scarcely a cloud bedimmed the sky. Floods of sunshine poured down to give life and animation to a scene which we witness but once in a decade.

'Early in the morning a merry peal was rung from the Cathedral Tower, which speedily awoke responses from the belfry of St Mary de Crypt and other churches in the city. Flags and streamers were hoisted above the roofs of the houses and banners were strung across the streets high across the crowds of holiday makers who poured into the city.'

A local committe had been formed and a guarantee fund was liberally subscribed. A list of agricultural prizes was offered, there was in addition a Flower Show and a public dinner. The band of the Grenadier Guards and the local volunteer bands were engaged.

There were nineteen classes for cattle; Herefords and Devons were judged together in one class and Shorthorns and other breeds in another. There were twenty classes of sheep, four of pigs, five of horses, best sack of white wheat, best sack of red wheat, three cheese, one wool, two sheepdogs and best collection of agricultural implements.

The reporter waxed lyrical over the flower show, which was held in two large tents, one for plants and one for cut flowers:

> 'The woodbine pale and wan —
> Hypericum all bloom, so thick a swarm
> Of flowers like flies clothing her slender rods,
> That scarce a leaf appears'.

He then went on to decry the fact that all our favourite flowers came from a foreign source.

While some were inspecting the cattle, or admiring or dancing to the band of the Engineers, the younger visitors were engaged in sports organised by Captain de Winton and members of the City Rifle Company.

The dinner took place in a large tent and was presided over by E. Holland MP, supported by the Mayor and his lady, Col Kingscote MP, R. S. Holford MP, Hon W. Bathurst, Hon C. F. Berkeley, W. Phillips Price, T. B. Lloyds Baker, T. Gambier Parry, the Dean of Waterford and many other gentlemen of note.

The Chairman rose and said they were limited for time because many of them had a train to catch. Nevertheless they found the time for fifteen toasts, all of which brought a suitable response.

'At night additional gaiety was given to the old city by several brilliant gas illuminations. In front of "The Tolsey", "Speed the Plough" appeared in large letters of flame, with a border of oak foliage, flanked by blazing stars and surmounted by a gigantic crown. Gas stars were also erected in Westgate Street and minor illuminations in other parts of the town.

'As the evening drew on several thousands of persons congregated. Dancing was of course the main amusement and at one time no less than a thousand were footing it to some lively quadrilles. The refreshment tents were well patronised and groups of people were scattered around discussing (sic) Porter and sandwiches.

'Great credit was given to the police for maintaining order. Several of the "light fingered gentry" appeared but their well known faces were recognised by plain clothes policemen at the railway station. They were given the option of returning or being put into custody.'

It was estimated that some 20,000 attended the show and admissions amounted to £825.

The results were so satisfactory that it was resolved that the next show should be held in Cheltenham. However, it was found that this would clash with the Royal at Worcester, so the 1863 show was held at Cirencester. This was accompanied by heavy rain; because of this and the expense of the flower show a loss was made.

This convinced the committee that, when choosing a site for a show, a larger guarantee fund must be requested. Therefore, when Cheltenham invited the Society in 1864, £600 was fixed, the Society calculating their expenses at £1,200. The Society paid the prize money, judges' expenses and costs of fitting up the cattle yard. The local committee met the costs of the local classes and flower tent.

The show was held in Pittville in the area round the Pump Room. The cattle were confined in pens at the back of the Pump Room and in front of the Spa, a platform was erected on which the Band of the Royal Marines and the Town Promenade Band played alternately.

There were upwards of fifty entries of Shorthorn cattle, the Hereford and Devon heifers were more numerous than the Shorthorn and there were six entries of dairy cows.

There was a good show of Cotswold sheep but a much smaller show of Southdowns, one Shropshire and a few Oxford Downs. There was a large show of pigs being mainly Berkshires and a few 'Improved Blacks'.

The entries included 'some able cart horses', hunters, hacks and a dozen ponies. A large machinery display was catalogued to include a liquid manure distributor, mangling machines, rollers, iron stable fittings, weighing machine and a portable steam engine with combined threshing machine.

There was a dog show held 'With a view to the enticement of the aristocratic classes. It was arranged so that the highest ladies in the land could wander through the exhibition and not see a thing to which their delicacy could take exception. Indeed the dog show was patronised as much by the ladies as by the rougher sex and they were by no means behindhand as exhibitors'.

1865 was the first year that the show visited Tewkesbury. Held in conjunction with the annual regatta on the River Severn it promised to be a grand affair. Unfortunately it co-incided with a general election and heavy rains on the second day put a lot of people off.

Among those present at the dinner were Mr J. R. Yorke, W. E. Dowdeswell, C. Yorke, C. Hayward and Rev Canon Davies. The latter, replying to the toast of the Bishop and clergy, threw in a heavy hint that support for the restoration of the Abbey would be appreciated.

Cattle plague and rinderpest caused the show to be cancelled in 1866 and in 1867 it returned to Cheltenham. The town guaranteed £600 — and there was an attendance of 20,000. 'No town in the county could fill such a site with an assembly of rank and fashion, youth and beauty, such as congregated in the noble Pump Room'.

The Show was held in conjunction with the Cheltenham Horticultural Society, which included grapes and hot-house plants. There was also a honey section. 'Bee keeping is a very useful adjunct to the cottager's income, its sale must come in useful to enable him to purchase his store pig and provide him with food for the winter.'

The dinner took place in a large tent near the horticultural section and was laid for five hundred. Only about half this number turned up and this was ascribed to the unpopularity of the Chairman, Sir George Jenkinson, 'with whom many positively refused to have intercourse'.

Among those who did turn up were Earl Bathurst, Mr E. Holland MP, Baron de Ferriers, Captain de Winton, Mr Engall and Rev Healy. Toasts were drunk to the Army and the preparations for the Abyssinian Expedition. The response to the Navy was by Captain Robertson, who had served in HMS *Victory* under Nelson.

The show included a section on fine arts and the Baron de Ferriers expressed the wish that farmers would become so prosperous that specimens of art might be purchased. A few years ago about the only pictures a farmhouse possessed was a gaudy print of George IV, and as to china nothing beyond an old willow pattern. He hoped the time would come when every tenant farmer would have his pictures and every tenant farmer's wife a cabinet in which to put her china.

Sadly the show results were not reported because 'of the marked discourtesy with which the press were treated by the secretary, Mr Trinder. He should realise the advantage which accrues to the individual and the society, by the cultivation of manners as well as by the cultivation of pigs and turnips'.

One way and another it would seem that relationships overall were not smooth. At the winding-up dinner on 9 October, it was reported that no cattle were exhibited because the two railway companies refused to co-operate. The joint show secretary, Mr Cochrane, in replying to the vote of thanks, said the Great Western Railway had run a special train — in the opposite direction — without much success. (Loud cheers). For all this he could report a clear profit of £130.

It was late when the party broke up, several good songs contributing to the harmony of the evening. The veteran vice chairman, Mr G. A. Williams, notwithstanding his threescore years and twelve, was one of the contributing causes of the evening's pleasure.

'He still remembers that he once was young,
His easy presence checked no decent joy'.

In 1872 the show returned to Gloucester, but heavy rains made it a miserable event. Cancellation was considered but, because some of the exhibits had arrived, it was decided to carry on, even though the ground was partially flooded.

In 1873 the Royal Show visited Stoke Park, Bristol and in 1874 there was a return to Tewkesbury. At the dinner J. R. Yorke MP remarked that Parliament had removed the tax on horses, which would be of great benefit to farmers. Sir E. Lechmere presided and commented that he looked forward to the day when Worcester and Gloucester would combine.

The Press reported that they were rigidly and not very courteously excluded from the ring, while others, having no official duty to perform, were allowed within the sacred enclosure, their only passport apparently being the favouritism that is not confined to 'kissing'.

Over the years, besides the towns already mentioned, the Society visited Dursley, Berkely, Moreton in Marsh and Bristol. Each of these tried to out-do each other in their decorations of main streets and their welcome. Sadly however, they either made a loss or at best broke even. So the show returned to either Gloucester or Cheltenham at regular intervals, where it was reasonably sure of recouping those losses.

By the eighteen eighties the Society had grown to such an extent that there were twenty-six classes for cattle, seventeen for sheep, ten for pigs, ten for hunters, hacks and ponies, ten for cart horses, four for jumping horses, five for cheese, seven for butter and one for cider. Usually prizes were for first and second only; eight pounds for large animals, with four pounds for second and five and three for pigs and sheep.

Herdsmen were given five shillings and half a crown but shepherds fared better in that they got three pounds and two pounds. There were also prizes for implements and six medals were awarded.

Special prizes were awarded for tenants in the Berkeley and Cotswold country. Individuals sponsored many of the classes and the prize money varied according to the generosity of the donor. Thus the prize money for tenants in the Berkeley country competing in the butter classes amounted to £16 but tenants in the Cotswold only got £10.

The 1888 show at Moreton was held in the Park belonging to A. B. Freeman Mitford CB, and was notable in that the distinction between tenants and landowners was abolished in the competitive classes. A new feature was a sale of stock on the showground by the well-known auctioneer John Thornton. Once more the railways came in for considerable criticism for their lack of co-operation and delivery of stock at odd hours during the night.

The Society went through a series of reverses at this time, partly as a result of inclement weather and partly because the smaller towns failed to provide sufficient income to cover expenses. A visit to Bristol in 1889 was a washout in more senses than one and the Society went into a state of 'suspended animation' for the next two years.

Suggestions for a possible resuscitation were made and meetings were held with a view to holding a show at Cheltenham, but the arrangements fell through. Then came an invitation from Stroud, which had previously hosted the show in 1884. The Stroud local committee provided a guarantee fund of £2,000 and £200 towards the prize fund. The Show was held in Stratford Park and proved to be such a success that all liabilities were discharged.

The Bath and West Show was held at Gloucester in 1893 and the Gloucestershire Society combined with that, for it was felt that there would not be room for two shows in one year.

In 1894 they visited Cheltenham and in 1895 went to Cirencester. HRH the Prince of Wales visited the show, for the celebraton of the jubilee of the Royal Agricultural College. His Royal Highness also exhibited in the yearling Shorthorn Bull class and came second to the champion.

The general financial depression of the times continued to threaten the future of the Society and for the remainder of its existence it rotated round Cheltenham, Gloucester, Stroud and Tewkesbury.

A special effort was made to put on a good show at Cheltenham in 1898. It was held in a sixteen acre field belonging to Mr Vassar Smith of Charlton Park, the grandstand capable of holding 2,000 people and the refreshment tent was 200ft long by 60ft wide.

Prize money amounted to £1,000 and it was generally acknowledged that the Shorthorns were the best that had been seen for many years. In all there were eighty classes for stock and horses and a shoeing competition was a new feature. There were also extensive poultry and produce classes. Added to the jumping and polo matches were the machinery displays and it was difficult to understand how it was all accommodated in sixteen acres.

In 1902 the show visited Tewkesbury where there was a record entry of 235 cattle. The event was somewhat overshadowed by the coronation celebration for King Edward VII. To make up for this it received a visit by the Crown Prince of Siam. He was met at the station and escorted to the showground, in Breakingstone Meadow, by a detachment of Volunteers under Lieutenant Moore. Tewkesbury locals were amazed that he spoke perfect English and looked like a normal human being. What he thought of the Theosburians is not recorded.

The Society continued holding successful shows in the main towns of the county. In 1906 they combined with the Bath and West Show at Swindon and in 1909 with the Royal Show at Gloucester.

In 1911, after seven years in other parts of the county they re-visited Gloucester. On this occasion they returned to the practice of having a three day show, the more recent shows having been two day events. This was not popular with exhibitors, because of the inconvenience of having men and exhibits away from home for a longer period.

'The Tuesday morning was ideal from the visitor's point of view, the sky being clear with no sign of rain, though no doubt many of the farmers present would have been glad to tramp through mud as deep as that which covered the ground at the time of the Royal's visit, if their crops could have had the benefit of the rain which they needed after the long drought'.

'The show was held on the Oxleaze on the site used for the Royal Show and included some of the ground made up by the corporation for that event. The area enclosed was nearly twenty acres and it was found possible to lay out a polo ground 900ft by 300ft in front of the grandstand. This ground was also used for the riding and jumping, the jumps being removed for the polo.

'The dry weather made the ground very hard but this was overcome by the Corporation's action in commissioning the fire float *Salamander* to water the ground from the river on Saturday and Monday evenings'.

The fifty-third show of the remodelled Gloucestershire Agricultural Society was held at Cheltenham on 17 and 18 June 1914. This was the thirteenth occasion the show was held in Cheltenham, this time at the Stonewall Fields on the Prestbury road, tenanted by Mr A. Wigget. The two previous shows of 1905 and 1910 were held on the same site.

In addition to cattle the show offered a variety of attractions. The band of the 18th Hussars was engaged for both days, there were horse shoeing, butter making, poultry trussing competitions and bee keeping demonstrations. There were parades of horses and cattle in the ring each day and jumping and driving competitions.

The exhibits of implement manufacturers were numerous as were those of Colonial Governments, who seized upon the occasion to bring the possibilities of the colonies before the British agriculturist, who in preference to any other class they wished to attract to other 'Britains beyond the seas'.

A great feature was the parade of turnouts, which left the Queens Hotel at two o'clock. It included heavy wagons sent by many tradesmen within a five mile radius of Cheltenham.

Mr M. H. Hicks-Beach was President and the attendance figures on the first day, as was usual at Cheltenham, were at 3,744 a record. A violent thunderstorm on the second day put some people off and they had to be satisfied with a modest 4,470.

Names still remembered in the county such as Garne, Hobbs, Cridlan and Withers were prominent among the stock exhibitors as was Mr W. N. Unwin in the hunter class.

At the annual general meeting which, as usual, was held on the second day, a deputation from Cirencester invited the Society to hold the 1915 show at that town, which was considered to be 'The Metropolis of Agriculture in the Cotswolds'. The invitation was accepted, Earl Bathurst was elected President and the show was to be held in Cirencester Park.

Sadly the 1914-18 war intervened and it had to be cancelled. The Society in fact never got going again after the war ended.

Although the Society and the countryside generally were run by the nobility on autocratic lines, everyone knew their place and what was expected of them. This fostered a spirit of mutual trust and understanding which cut right across class. This spirit of trust was carried over into the Royal Gloucestershire Hussars Yeomanry, which many of the younger nobility and yeoman farmers joined at the outbreak of war. The Regiment saw service in Gallipoli, the Sinai desert and up into Palestine. Their casualties were heavy and there can be no doubt that the loss of this flower of the county was one of the deciding factors in the failure of the Gloucestershire Agricultural Society to be revived after the war.

		£	s	d
1.	For the best Dairy Cow, more than 4 years old, in Milk or in Calf, having been the property of the exhibitor at least 12 months	8	0	0
2.	For the second best ditto	4	0	0
3.	For the two best Dairy Heifers in Calf, bred by and the property of the exhibitor	5	0	0
4.	For the two second best ditto	3	0	0
5.	For the two best Dairy Heifers, under two years old, bred by and the property of the exhibitor	5	0	0
6.	For the best Bull, of any breed, of more than three years old	10	0	0
7.	For the best Bull for Dairy purposes	10	0	0

Premium, the gift of T. J. L. Baker, Esq.
For the Bull, Cow, and produce, which shall be declared by the Judges the best for dairy purposes, thorough-bred; or as nearly so as possible; the cow to have been in the possession of the exhibitor when put to the bull, and to be so with the produce at the time of being exhibited. The bull not necessarily to be the property of the exhibitor, or of any Member of the Association. No animal to be incapable of gaining this premium on account of its gaining or having gained any other.

N.B. The above Stock to be fed on Hay, Roots, Grass, and other Green Vegetables only, and so certified.

FAT CATTLE.

		£	s	d
8.	For the best Fat Ox, fed in any way by the exhibitor	10	0	0
	If bred also by the exhibitor, additional	2	0	0
9.	For the best Fat Cow, as No. 8	8	0	0
	Additional, as No. 8	2	0	0
10.	For the best Fat Steer, under four years old, as per regulation No. 8	4	0	0
	Additional, as No. 8	1	0	0

Premium, the gift of J. W. Walters, Esq.
For the best Fat Heifer, having produced a Calf 10 0 0

25.	To the exhibitor of the best skim Plough, applicable to stiff land, his own invention or improvement	2	0	0
26.	To the exhibitor of any Tool or set of Tools, Implements, or Machine, entirely or materially new, found to be more useful in any branch of Agriculture, Arts, or Manufactures, than any before known for similar purposes, a premium in proportion to merit according to the funds.			

LABOURERS.

27.	To the Labourer, or Widow of a Labourer, who shall be recommended by a Subscriber, who has brought up or is now supporting the largest number of legitimate children without assistance from any parish, of which facts, requisite certificates must be produced	2	0	0
28.	To the second ditto	1	0	0

MANUFACTURERS.

29. To the exhibitor of the best piece of Woollen Cloth made wholly from British Wool.

PLOUGHING MATCHES.

30. To the owner of any Plough drawn by not more than four Horses, which shall best

		£	s	d
11.	For the best three Breeding Ewes, long wool, having each bred up a Lamb within the present year, and under 33 months old	3	0	0
12.	For the best three ditto, short wool	3	0	0
13.	For the best three Sheerhogs, long wool	3	0	0
14.	For the best three ditto, short wool	3	0	0
15.	For the best 3 breeding Theaves, long wool	3	0	0
16.	For the best three ditto, short wool	3	0	0

Nos. 11, 12, 15 and 16, to be fed on Green Vegetable Food, Hay, and Roots only.

PIGS.

17.	For the best Boar	2	0	0
18.	For the best breeding Sow	2	0	0
19.	For the best Fat Pig, not being a Boar Stag, fed by the exhibitor, and at least 3 months previously in his possession	2	0	0

HORSES.

20.	For the best Cart Stallion, in possession of the exhibitor, and at least twelve months	5	0	0
21.	For the best Mare, and Foal her own offspring, in reference to the general purposes of agriculture, possession as No. 20	5	0	0

CHEESE.

22.	To the person who shall exhibit the best hundred weight of Cheese, made from land in his own occupation	3	0	0
23.	To the person who shall exhibit the second best ditto	2	0	0

Premium, the gift of W. L. Lawrence, Esq.
For the best Practical Treatise on the most eligible method of Tillage applicable to Clay Lands, so as to lessen the expences of cultivation, and enable the occupier, to keep a larger quantity of Stock 10 0 0

MECHANICS.

24.	To the exhibitor of the most generally useful and effective Plough, for stiff land, requiring the least power of draught, his own invention or improvement	2	0	0

plough half an acre of land within a given time, which shall be hereafter specified..

		3	0	0
31.	To the manager of ditto	1	0	0
32.	To the owner of the second best	2	0	0
33.	To the manager of ditto	0	10	0

The ploughing premiums to be given once in the year at a Ploughing Match, the time of which shall be fixed by the Committee of Management.

No premium to be awarded to any exhibitor who is not at the time a member of the Association, but the Committee recommend that encouragement shall be given to the exhibition of Stock and Implements by all persons, not members, who may wish to do so.

N.B. *No Animal to be allowed to gain two of the Society's Premiums at the same exhibition.*

At this Meeting Mr. W. L. Lawrence gave notice that there would be laid before the next general Meeting, for its sanction and support, a Petition to both Houses of Parliament, praying for the continuance of the existing Corn Laws, and for the general relief of the Agriculturists of this Country, from the oppressive burthens borne by them in direct and indirect taxation, and particularly in their present unequal competition with the untaxed Irish grower.

Schedule of classes for the 1834 Gloucestershire Show and minutes of a General Meeting (from two separate pages of a report of 22 February).

LEFT: Extract from the Minutes of 24 February 1838. RIGHT: Extract from Minutes in 1836. BELOW: Gloucestershire Agricultural Society Show was held at Tewkesbury in 1865, in conjunction with the Tewkesbury Regatta.

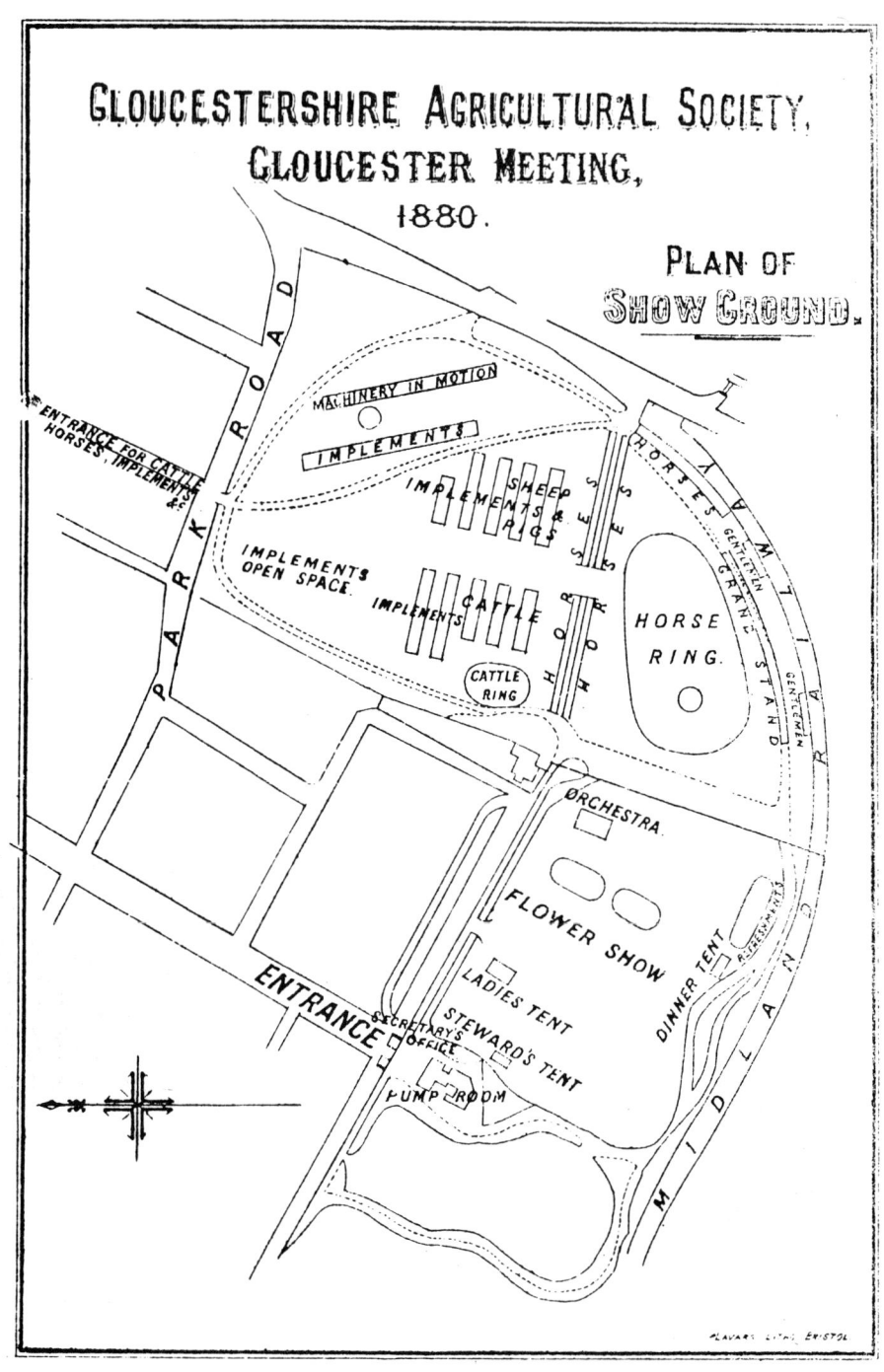

ABOVE: Site plan of Gloucester Show in 1880. OPPOSITE: Mr W. Garne of Northleach was a great supporter; his Cotswold ram won first prize; CENTRE: Mr Hobbs of Maiseyhampton exhibited this Oxford Down ram, which won first in its class. BELOW: Alderman C. W. Poole tells the tale to the showgoers.

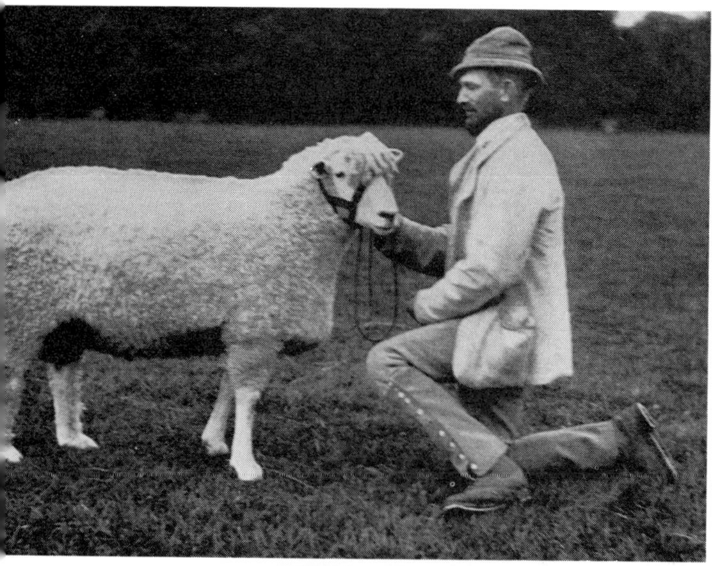

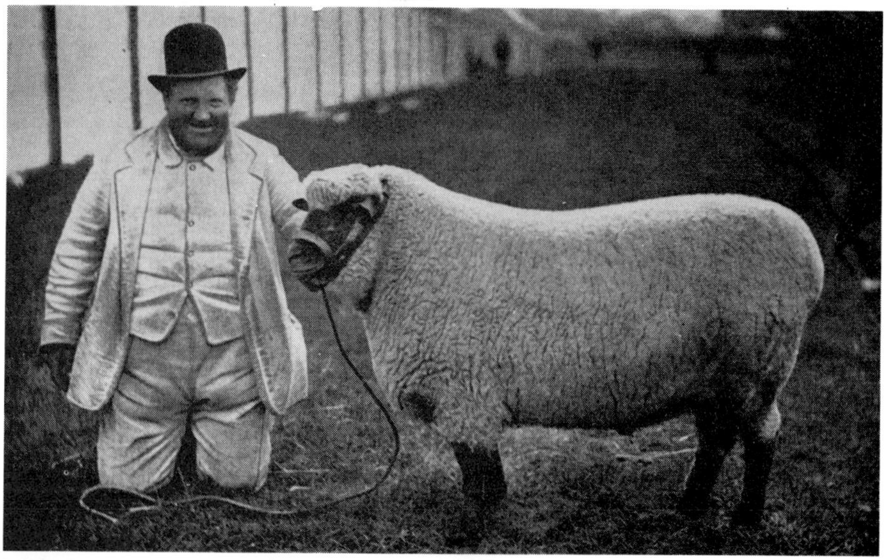

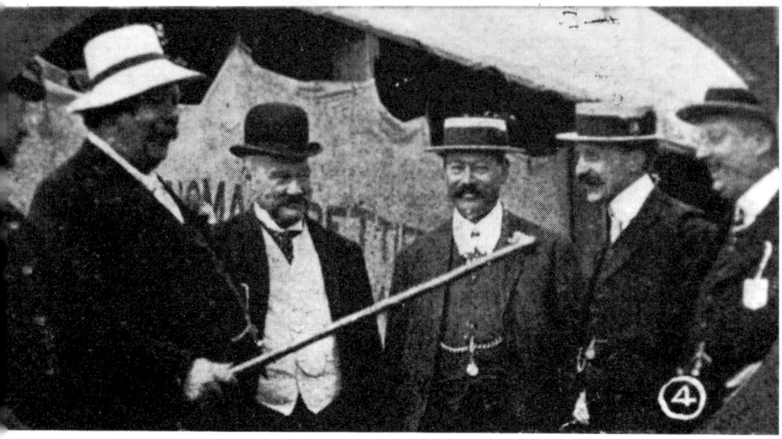

1883.
GLOUCESTERSHIRE AGRICULTURAL SOCIETY.

STATEMENT OF ACCOUNTS.

BERKELEY SHOW ACCOUNT.

Dr.		£	s.	d.
July, 1883.				
To Cash received for Admission to Show		639	19	0
,, Hire of Ground for Refreshments, &c		70	0	0
,, Entrance to Grand Stand		57	13	0
,, Fencing		27	12	10
		£795	4	10

Cr.		£	s.	d.
July, 1883.				
By Cash paid for Advertising		39	8	0
,, Printing		26	11	5
,, Bill Posting, &c.		12	1	0
,, Dinner Tickets for Reporters		1	10	0
,, Bryant and Garrick for Caps for Yard Men		1	16	0
,, Hungarian Band		63	0	0
,, Ticket Collectors, Gate Keepers, &c.		13	14	6
,, Saunders & Sons, for Fencing, &c.		67	1	2
,, ,, erecting Grand Stand		49	5	0
,, Bossom, for Hire of Chairs		2	0	0
,, Fear, for Water Casks		4	0	0
,, Organ, for Hauling Gorse		1	0	0
,, Extra Police Account—Allowance and Travelling Expenses		16	1	11
,, Sundry Payments		5	17	11
Balance		491	17	11
		£795	4	10

Gloucestershire Agricultural Society.

GLOUCESTER MEETING.

Catalogue of the Entries
OF
CATTLE, SHEEP, PIGS,

Horses, Cheese, Implements, &c.,

For the Premiums offered by this Society at their Meeting held at Gloucester, on the

27th, 28th, and 29th JULY, 1880.

JUDGES.

SHORTHORNS.
Mr. EDWARD BOWLY, Siddington House, Cirencester.
Mr. W. FAULKNER, Rothersthorp, Northampton.

HEREFORDS AND CHANNEL ISLANDS.
Mr. HENRY MIDDLETON, St. Frideswide, Cuttslowe, Oxford.
Mr. JOHN PRICE, Court House, Pembridge.

SHEEP AND PIGS.
Mr. JOHN GAY ATTWATER, Britford, Salisbury.
Mr. THOMAS PORTER, Baunton, Cirencester.
Mr. J. TREADWELL, Upper Winchenden, Aylesbury.

CART HORSES.
Mr. ROBERT CRADDOCK, Lyneham, Chipping Norton.
Mr. GEORGE ROBINSON, Slimbridge, near Stonehouse.

HUNTERS.
The Right Hon. EARL COVENTRY, Croome Abbey, near Worcester.
Mr. C. A. R. HOARE, Cirencester.

HACKS AND PONIES.
Major PROBYN, Huntley Manor, near Gloucester.
Captain A. HOLME SUMNER, Rosehaugh, Cheltenham.

CHEESE.
Mr. B. BRUNSDON, Ross.
Mr. EDWARD BRETHERTON, Grosvenor House, Cheltenham.

The Prizes are distinguished thus:
FIRST PRIZE .... Red Rosette.	RESERVE NUMBER, Green Rosette.
SECOND PRIZE ...... White ,,	CHAMPION PRIZE... Red and White Rosette.
THIRD PRIZE .... Blue ,,	RESERVE NUMBER, Green and White ,,

ED. TRINDER, Secretary, Cirencester.

PRICE ONE SHILLING.

GLOUCESTER:
"JOURNAL" STEAM PRINTING WORKS, WESTGATE STREET.

Gloucestershire Agricultural Society

ESTABLISHED 1829.

GLOUCESTER SHOW,

TUESDAY, WEDNESDAY & THURSDAY, July 25, 26 & 27, 1911

President - HIS GRACE THE DUKE OF BEAUFORT.

PROGRAMME

ADMISSION—

1st Day, 2/6 ; 2nd and 3rd Days, 1/-. Grand Stand, 1/-

Grand Stand Tickets are available for the whole of each day, but not transferable.

ROBERT ANDERSON,
Cirencester.   Secretary.

FIRST DAY, Tuesday, July 25th.

ADMISSION 2/6. :: Open from 9.30 a.m. to 7 p.m.

9.30 a.m.—The Show Yard opens. Judging of Cattle, Sheep, Pigs, Horses, and Dairy etc. Produce.
11.30 to 1.30.—The Band of the Royal Gloucestershire Hussars. For programme see following pages.
1 p.m.—Public Luncheon.
2 p.m.—Working Dairy Competition, Class 1.
2.30 p.m.—Parade of Horses (all Classes) in the Ring, commencing with Class 51.
3 p.m.—Judging of Jumping and Driving will commence with Class 104, followed by Classes 101, 103, 98 and 96.
3.30 to 5.30 p.m.—The Band of the Royal Gloucestershire Hussars.
3.30 p.m.—Bee Demonstration by Mr. E. J. Burtt.
5 p.m.—Poultry Lecture and Demonstration in Working Dairy by Mr. H. R. Howman.
5 p.m.—Polo Match in front of Grand Stand—Tigers v. Kemble.
7 p.m.—The Show Yard closes.

ABOVE: Accounts for the Berkeley show in 1883. LEFT: Judges' list for the 1880 and RIGHT: part of the programme for the 1911 Gloucester Shows.

68

LEFT: Sir James Bruton, Mayor of Gloucester in 1911. Four generations of his family were connected with the Society. RIGHT: H. Dent-Brocklehurst, President, talking to Mr R. Anderson, the Show Secretary in 1911. CENTRE: Messrs Cuss, R. W. Hobbs, T. James, Hodges and S. Kite at Gloucester Show, 1914, held in Cheltenham. BELOW: Parade of 'Turnouts' leaving the Queens Hotel, Cheltenham, 1914.

The first show of the combined Three Counties held at Gloucester in 1922.

SHOWTIME! —
The Three Counties Agricultural Society

At a Council meeting held on the second day of the Hereford and Worcester Society show, on 8 June 1921, a letter was received from Mr L. C. Wrigley, secretary of the Gloucestershire Agricultural Society, suggesting an amalgamation. The suggestion was favourably received by the Council and a committee was appointed to represent the Society in the negotiations.

The committee subsequently recommended that the two societies amalgamate under the title of 'The Three Counties Agricultural Society'.

It was further agreed that the show be held in the three counties in turn and that the 1922 show be held in Gloucester, the Council be formed by 21 members from Hereford, 14 from Worcester and 14 from Gloucester.

Added to the Council would be the President, who should reside in the County in which the show was to be held, three Vice Presidents, one from each county and the Mayor of the town which the show visited.

A guarantee fund was formed into which the Hereford and Worcester Society put £2,000 and the Gloucester Society £750, this being Gloucester's total assets.

These arrangements were agreed by the Council and members of both Societies and the whole amalgamation was carried through most amicably.

On 31 December 1921 the Hereford and Worcester Society was formally wound up. The surplus assets were invested and the income divided in equal shares and disposed of by a Committee of members from each county.

The 1922 and the first Three Counties Show, was held at Gloucester under the Presidency of Colonel Russell Kerr. A report in the *Cheltenham Chronicle* stated that 'The showground was an ideal one and everything was admirably arranged. This fact combined with the magnificent entries of stock led many of the experienced showmen to place it second only to the Royal Show'.

The following year the show was held at Malvern to avoid clashing with the Three Choirs Festival. It was at this show that the local committee, having a £20 surplus on their prize fund, proposed that the money be used for a herdsmen's supper. This was held on the second evening of the show and started a feature that has been continued to the present day.

The 1924 show was held at Worcester and the Council had the choice of two sites — Perdiswell Park and Pitchcroft racecourse. Despite difficulties of fitting in the show so as not to obstruct race meetings, Pitchcroft was preferred and this turned out to be an unfortunate choice.

At an emergency meeting of the Council on Monday 2 June 1924, the Secretary reported that he had been woken up early the previous day (Sunday) and informed that the showground was under water. The only way he could inspect the showground was by horse and cart. There was already some livestock on site and these were immediately moved to high ground and found accommodation in various stables in the city. The Secretary tried to get in touch with the showyard director but he was cut off by the flood.

By Monday morning the showground was covered in eight feet of water. The Mayors of the cities were informed of the position and the railway companies were requested to stop all trains coming to the show.

After considerable discussion and considering alternative dates, the Council had no alternative but to cancel.

Members and exhibitors supported the Society generously; a loss of £2,144 was sustained, which could have been a great deal more. To try to alleviate the position, the finance committee discussed the possible introduction of Life Membership subscription, but they eventually decided against.

Although till then the show was held some time in June, no definite date had been fixed. After the Secretary had consulted with other show secretaries it was decided to make it known that in future the show would be held on Tuesday, Wednesday and Thursday during the second week of June. From that time, June 1925, these dates have been rigidly held, except in 1947, when the show was postponed for one month, because of foot and mouth disease.

At the March meeting in 1926 it was decided to have a rural industries section and also to provide a tent for the Women's Institute. It was also reported that Worcester City Corporation was in process of negotiating the purchase of Boughton Park for a future showyard. The negotiations fell through and the committee was left with having to chose between Perdiswell Park and the racecourse. After the 1924 flood they did not waste much time in deciding on Perdiswell.

In his report for 1927 the Secretary stated that, notwithstanding the depression and losses sustained at most shows during the year, the Society had cash and other assets of £5,017.

The Gloucester show resulted in a profit of £988 and the total cash prizes amounted to £2,739. The general opinion from the machinery exhibitors was that, considering the depression, business was as good as could be expected.

Considerable time was spent lobbying MPs and co-operating with other societies in an attempt to get entertainment tax abolished on military displays. £25 was spent on a joint appeal with other shows to abolish income tax on show receipts.

The 1929 show at Gloucester was honoured to have the Duke of Gloucester as President. Unfortunately His Royal Highness could not attend, as he was in Japan and HRH Prince George, later the Duke of Kent, deputised.

At the 1930 show, after being pressed by the Ministry of Agriculture, it was agreed to have separate accommodation for TT and non-tested cattle.

Between the wars the sites for the shows, with one or two exceptions, rotated round Hereford racecourse, Perdiswell Park, Worcester and the Oxlease, Gloucester. Money was hard to come by during the depression years. The local committees, who used to arrange an entertainment on the Thursday evening of the show, often found it difficult to raise the necessary amount.

At around this time (1930) it was suggested that the Society purchase a permanent site in each of the three counties, to overcome the recurring difficulties, but this came to nothing.

One of the first buildings to be erected on the showground was a sectional structure, used as a site office, and the 1931 show saw the first appearance of such a building. When the Society moved to its present permanent home at Malvern, it was refurbished and used as the President's pavilion, until the present permanent structure was incorporated in the Severn Hall.

The 1932 show was held at Gloucester and on 24 May an emergency meeting of the Showyard Committee was called. Due to heavy rains the showground was flooded to a depth of two feet. The meeting was adjourned to 27 May and again to Sunday 29 May. The Showyard Committee decided to carry on but, at a combined meeting of the showyard and finance committees it was decided to abandon the show for that year. An emergency meeting of Council was called for the following day. After hearing a report from the city surveyor that the flood was dropping and the ground — aided by the Fire Brigade — was drying, it was decided after all to carry on.

The show was held on 7, 8, 9 June and, after great efforts by the local committee and the City Council, the showground was cleared of water.

Because of this near disaster several other sites were considered for the 1935 show but, after Mr Scudamore, the city surveyor, had stated he could raise the level of the Oxlease using city refuse and afterwards turfing over, it was agreed to carry on using this site.

Because the Oxlease might not be ready by 1935 and in view of the fact that Cheltenham was anxious to host the show, its invitation was considered, but it was not pursued.

The Duke and Duchess of York visited the Hereford show in 1934 and stayed for three hours and there was a record attendance.

From the minutes of this period it is clear that the Mayor, Town Clerk, and others were very much involved in helping towards the success of a show, when it visited their particular city. In many instances they took part in the various committees, site meetings and general arrangements.

Much time seems to have been taken up at finance meetings interviewing various contractors, trying to persuade them to reduce their bills. This they usually did, but it happened with such regularity that it poses the question whether they added a bit on in the first place, because they knew this would happen.

In 1935 the show was visited by the Duke of Gloucester.

There was considerable discussion in the mid-thirties concerning the constitution of the Society. The Society was originally formed by the passing of a resolution at a meeting of members. In the event of the Society running into financial trouble this could put the members, especially Council, in some difficulty. It was suggested that the Society be formed into a limited company, but this was not pursued. The situation has more recently been resolved and members can sleep easy inasmuch as they would not be liable in the event of disaster.

For the 1937 show a wooden pavilion 70ft by 30ft, with a floor raised two feet from ground level, was provided for the use of members. Previously they had had to make do with a tent.

There was considerable feeling about the larger societies, such as the Bath and West and Royal Counties Shows, encroaching on the territory of shows such as the Three Counties. This came to a head when the Bath and West accepted an invitation to visit Cheltenham in 1940 or '41.

They suggested that the Three Counties join with them for that year, or that they give their approval. Council felt they could do neither and there was considerable ill-feeling. The issue drifted on for some time and became steadily more acrimonious. One difficulty was that the Agricultural Engineers Association felt they could not support both. The feeling was that if the show was held it should be on a fifty-fifty basis. The Council comment was that the Bath and West appeared to want the Three Counties to join in so they could walk off with all the profits. In the event the war intervened and the Bath and West was not held until after the war.

There was of course no activity during the 1939-45 war. The Council on 22 December 1939 appointed an emergency committee consisting of the Finance Committee and the showyard directors, giving them powers to take such action as necessary to keep the Society afloat.

There were regular meetings of the emergency committee and Council throughout the war. In October 1942 it was reported that, although it had been agreed that members' subscriptions be waived fur the duration, many members had still paid. This income, plus the income from investments, enabled the Secretary to report a cash surplus of £229 18s 9d. Out of this grants were made to the Red Cross, the Women's Land Army Benevolent Fund and an RSPCA fund to supply veterinary supplies to Russian animals!

Council members attending the meeting were Ald J. W. Hewitt, deputising as President for Lord Somers, Messrs W. G. C. Britten, William Smith, G. F. B. Witcombe, L. Joseland, A. H. Chew, A. Baldwin, J. Bellamy, G. H. Bray, P. E. Bradstock, G. H. Edwards, A. Allsbrook, Stafford Weston, J. P. Terry and W. H. Homes. Also present were Mr T. A. Matthews, solicitor and Mr T. H. Edwards, the Secretary.

Although the ending of the war was far away, the Council nevertheless spent some time discussing starting up again. It was left to the executive committee to get things moving as soon as conditions allowed.

ABOVE & OPPOSITE: Pre-war Show scenes.

BELOW: The Show in 1948.

The flooded showground in 1924 on Pitchcroft Racecourse, Worcester.

ABOVE: Cattle parade passing the grandstand; CENTRE: Hereford cattle parading and BELOW: the stockmen, without whose devotion to their animals the Show could not function.

ABOVE: A good level class of Hampshires before the judge. LEFT: Mr T. H. Edwards, Secretary of the Show from 1902-1946 inspecting it on his pony. RIGHT: Visit of their Royal Highnesses the Duke and Duchess of Gloucester in 1954.

78

ABOVE: Display of craftwork by a Romany family in the educational section. BELOW: Visit of a Nigerian Government delegation.

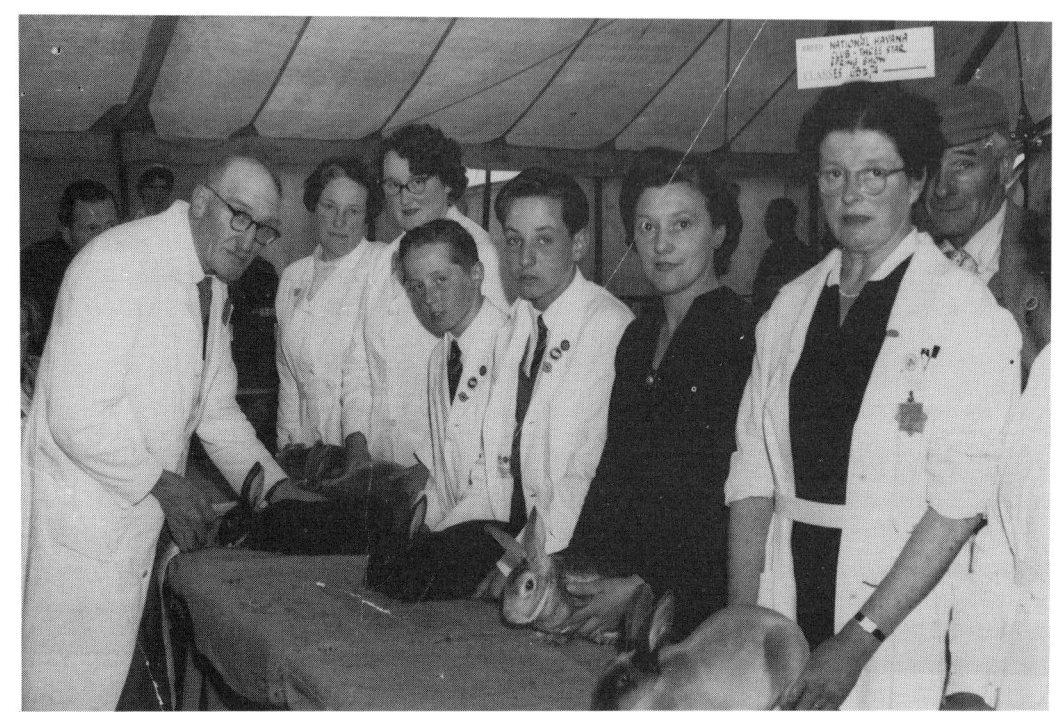

ABOVE: Judging the Rabbit Section. BELOW: Italian flag-throwing display.

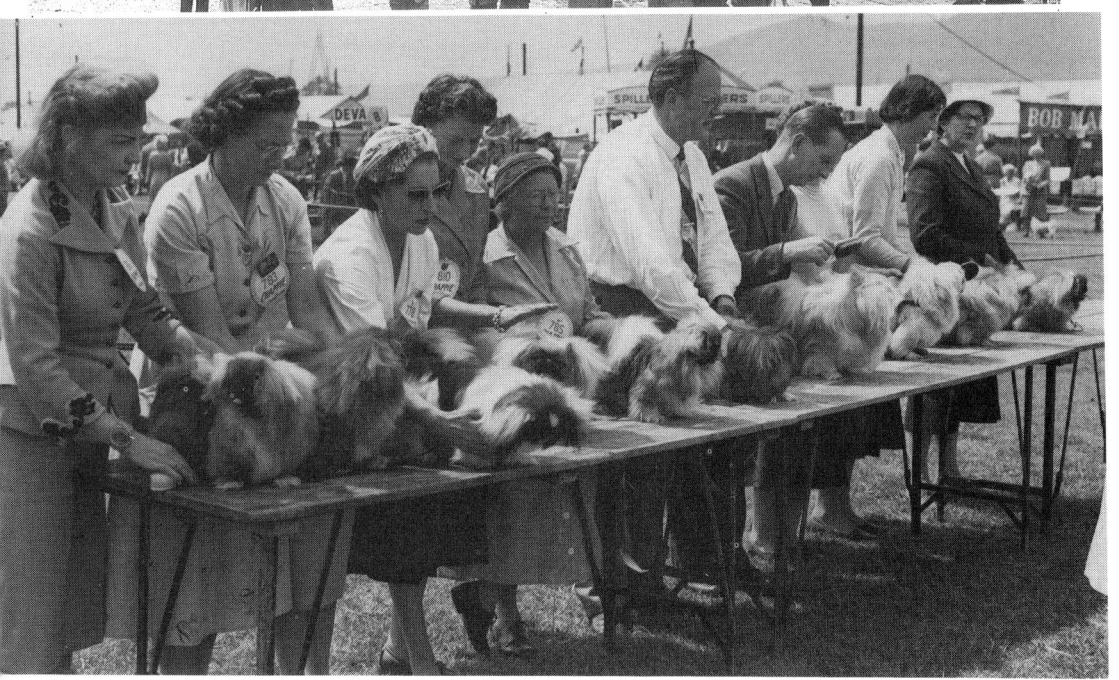

ABOVE: The stand of H. Burlingham & Co of Evesham, Staverton 1948. The implements displayed may be compared with those of more modern times. BELOW: The Three Counties Championship dog show section has long been a main feature of the June show.

LEFT: The Society is indebted to voluntary bodies, such as the British Red Cross, who give their services at the Show. RIGHT: The Young Farmers' Club section is a thriving feature of the Show. Here is a typical exhibit. BELOW: Agricultural Workers' long service award winners.

Home Sweet Home —
The Society at Malvern

With the cessation of hostilities in 1945 it was hoped to hold a show a year later. Shortage of materials for the erection of the showyard, and reluctance of the Ministry of Supply to allow scarce resources of labour and material for this purpose, forced the Council to decide that rather than risk a flop in 1946 they would go all out to put on a good show in 1947.

At the Council meeting in December 1945 Mr T. H. Edwards felt obliged to give way to a younger man and offer his resignation as Secretary after fifty years in post. Mr Glynne Hastings, who for some years had run a successful show at Thame, was the successful candidate out of 700 applicants.

The Council presented a cheque to Mr Edwards on his retirement and an illuminated address, which was executed by Mr Hastings.

At the Council meeting on 30 May 1946 it was agreed to purchase Berrington House, King Street, Hereford, for use as the Society's office. Previously the office had been at the Corn Exchange, Leominster.

It was a wonder that the 1947 show got going at all; few shows have been so beset with difficulties. Perdiswell Park at Worcester had been turned into allotments, the old Hereford ground had been ploughed up and the Oxlease in Gloucester was covered in pylons.

A suitable site was eventually found at Hampton Bishop in Herefordshire. Because of post-war shortages, there were not enough poles to bring electricity to the showground and they had to use mobile generators. There was only enough piping to bring water to the gates and the Fire Service was called on to distribute it round the ground. To cap it all there was an outbreak of foot and mouth disease, two miles from the showground, just before it was due to open on 17 June. After several alternatives were considered it was decided to postpone the show until July.

Because of commitments to other shows there was not enough timber for the grandstand and stock shedding. By superhuman efforts a grandstand was transported from the Royal Norfolk show, our own grandstand having gone to the Royal Cornwall.

To overcome the shedding problem the livestock was kept on the ground only one day. Cattle and hunters were exhibited on the first day, pigs, sheep and heavy horses second and riding ponies on the third.

Despite all the difficulties and forebodings of disaster, not least the difficulties of machinery exhibitors being able to get something to exhibit, there was a record attendance of 39,433 and a cash surplus of £2,853.

The 1948 show was held at Staverton Airport, Gloucester, the Oxlease being unavailable. It was held under the Presidency of Vicount Bledisloe and there was a record attendance of 63,422, which was not broken until 1974.

Prior to the show some considerable concern had been expressed about the scarcity of petrol allowing visitors to be able to use their cars to get there. Even though the war had ended there was still an extreme shortage. Mr G. N. Bruton pointed out that there should be no problem because Gloucester and Cheltenham were each only a fifteen minute bicycle ride away. Used as we are these days to jump in the car to travel the shortest distance, it is difficult to appreciate the lengths to which people went to get to a show after eight years' deprivation and austere conditions.

The 1949 show was held at Worcester and overseas visitors' facilities were introduced for the first time.

On 16 May 1949, at an informal meeting of directors of the showyard and other members of Council, it was agreed that, because of the difficulty of getting suitable sites, permanent facilities should be considered. At the Council meeting on 31 May 1949 a committee was set up to consider future policy, especially the future location of Society shows.

Because of the difficulties that arose since part of Hereford racecourse had been turned into an aerodrome, the 1950 show was held at Leominster.

Ever since the amalgamation of the three counties into one Society, Hereford as the senior had twenty-one members on the Council against fifteen each for Worcester and Gloucester. After some discussion on whether representation should be based on the number of members per county it was finally agreed that, as it was now effectively one Society, each county should be equally represented by eighteen representatives, one third to retire annually, and being eligible for re-election if they so desired.

This was put as a resolution to the general meeting of members held at Gloucester on 13 June 1951. It was also agreed that the President, past Presidents and the Mayor of the town in which the show was to be held, should be *ex-offiocio* members of Council.

During the nineteen fifties it became increasingly difficult to find sites in the three counties. With farming going through a period of prosperity, no-one wanted the disruption the show caused, certainly for two years and often for three. In addition the show was growing and the costs of putting up temporary facilities every year were escalating.

Used as we are to permanent facilities it is not easy to visualise the annual problems. So that the organisers had some idea where they were going, sites had to be booked up at least two years ahead. When a suitable site was found a layout had to be decided. This involved the Secretary and showyard committee walking the ground and making various suggestions. The Secretary then went off and drew up a plan, which the showyard committee then approved or amended.

To accommodate the car parks and showground proper a site of between a hundred and fifty and two hundred acres was required. It had to be reasonably level and easy of access to traffic. In earlier years availability of a railway station nearby was a necessity.

Since time immemorial Messrs L. H. Woodhouse & Co have been the main erectors and suppliers of temporary shedding, not only for the Three Counties Show but for many others throughout the country. There had to be close liaison with them, Messrs Woodhouse no doubt having their own problems trying to satisfy everyone.

The Society had a number of temporary buildings such as the showyard office, members' pavilion, furniture and other odds and ends, which they owned themselves. After each show this was all stored, temporary buildings dismantled and all packed into some large furniture vans which were parked in the yard at Berrington House. It was the pious wish of all concerned that what was thought to have gone into the vans came out again the following year.

After the show there was the clearing up and where necessary the reinstatement of the ground, to the owners' satisfaction. After the 1954 show at Staverton Airport, Gloucester, when the rains turned the ground into a sea of mud, it was several years before a settlement was agreed.

It was after this Gloucester show that the Council decided on 20 December 1954 that the search for a permanent site must be accelerated.

The committee inspected a number of possibilities and had almost decided upon a site at Elmstone Hardwicke, near Cheltenham. However, after an impassioned appeal from the Mayor of Hereford to Council, it was finally agreed to settle at Firs Farm, Malvern as being in every way more suitable. There is no doubt that, from its central position, ease of access, free draining soil and scenic beauty, the Society could not have made a better choice. The decision to acquire the site was finally ratified at a Council meeting on 20 April 1956.

The last travelling show was held at the Gloucester College of Agriculture at Hartpury. The first show on the permanent site was held in 1958, and was honoured with a visit by Her Majesty Queen Elizabeth, the Queen Mother.

In 1968 the Society was further honoured with a visit by Her Majesty the Queen, who spent four hours on the showground.

At a meeting of the finance committee on 14 August 1956 the Secretary reported that he had engaged as office boy a young lad called Lyn Downes. He commented that he would like to engage him permanently, while he was waiting to be called up to do his National Service in the Royal Navy. After that, Mr Downes returned to the Society and became assistant secretary.

By Mr Hastings' retirement in July 1972 the Council had realised Mr Downes' abilities and was unanimous in offering him the post of Secretary, later raised to that of Chief Executive. His ability was obviously appreciated in other quarters, notably in the Show Secretaries' Association. In the New Year's honours list for 1992 he was appointed MBE. This, besides being a great honour for Mr Downes personally, was also an honour for the Society.

In 1956, others on the staff were Mr David Evans, the assistant secretary, who afterwards became secretary to the Royal Counties Agricultural Society, Mr W. H. Wresthall the cashier, and Mr J. W. White.

The first five year programme on the new site included the laying of permanent roadways, installation of water, electricity and other services; also the building of a permanent office block and Council room.

Toilet blocks, shower rooms and an extensive tree planting scheme followed. 1974 saw the completion of the 56,000 sq ft cattle building, followed in 1976 by the Amenity Hall and in 1980 the 27,000 sq ft sheep building.

In addition the Society has purchased adjacent blocks of land as they became available, starting with fifty acres of woodland in 1977, which has been managed and improved to form an attractive campsite, in 1982 twenty acres of land at the rear of the grandstand for improved car parking and in 1984 the purchase of Langdale House and a further eight acres of woodland, giving the Society the ownership of the whole of the woodland fronting Blackmore Park Road. In 1988 the Society purchased Warren Farm plus thirty acres of land adjacent to the showground, and also in 1988 a further sixty acres of land adjacent to Langdale Wood. Langdale House, Warren Farm and its outbuildings were subsequently sold off.

The National Sheep Association purchased the Firs farmhouse and farm buildings for their headquarters.

The Three Counties Show is now only one of many events held on the Showground during the year and the permanent staff are fully engaged with the organisation of, for example, the Spring Garden Show, national sheep events, horse and pony shows, dog shows, car events and Caravan and Camping Club rallies.

In addition to the Three Counties Dog Section, which is a major three-day championship event held in conjunction with the Agricultural Show, the Society also hosts the West of England and Ladies Kennel Club Show in the spring, this being the largest outdoor dog show in the country.

In 1986 HRH the Princess Royal attended the show and opened the Severn Hall, a 24,000 square foot complex, fully heated and with kitchens to the highest standard. This means that the Society can accommodate events and functions all the year round.

The Society has become far removed from its original concept in that it has become a business in its own right. For instance, there are twelve permanent staff employed, plus a further six for six months of the year. In addition, three hundred and twenty temporary staff are put on for between three and ten days for the June show.

At least two thousand persons live on the showground for the whole of the three days. Many of them bring caravans or sleep in horse boxes, but there are many more who have to be found accommodation. No longer are herdsmen and owners prepared to sleep on a bundle of straw alongside the bull; they want and expect something better. Where not so long ago they used to wash in a bucket of cold water, or not at all, they now quite rightly expect to find hot running water and showers available.

Feeding the stock attendants, members and visitors is another consideration. When the author first exhibited in the immediate post-war years, he cooked his meals on a spirit or portable gas stove and other stockmen did the same. All this took place alongside the stock in the stock lines and it was a wonder the whole place was never set alight. Any shortfall was made up with tea and buns in the YMCA tent or by the hospitality of tradestands, who served similar fare.

Today it is estimated that well in excess of five thousand set meals are served each day of the June show. This takes no account of the various snack bars and refreshment served in the numerous tradestands throughout the show.

This has all meant that the kitchens, electrical fittings throughout the showground, washing facilities and general safety have had to conform and comply with the requirements of the several Health and Safety regulations. Years ago if someone got knocked over by a cow going into the parade ring or put their hand into a machine that was being shown, it was their own silly fault. Now things are different and the Society has to be constantly on the alert to ensure that regulations are complied with.

LEFT: The Society office, Berrington House, Hereford, dressed overall for HM the Queen's coronation. RIGHT: The President, the Viscount Bledisloe, opening the 1948 Show held at Staverton airport. BELOW: Spetchley Showground, 1949.

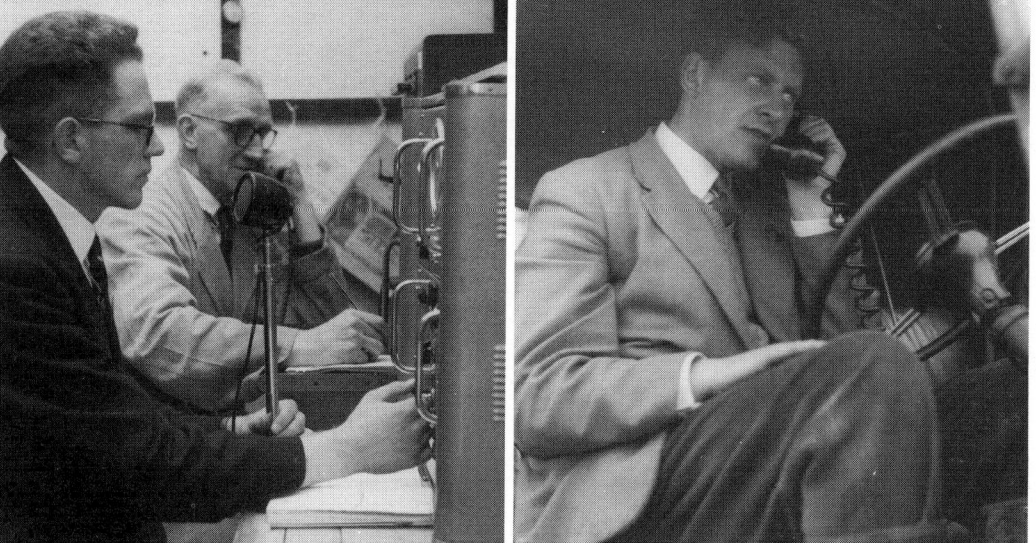

ABOVE: The Mayoral Procession from Cheltenham, processing to the opening ceremony. LEFT: Manning the communications centre. RIGHT: David Evans, sometime assistant secretary, receiving 'loud and clear'.

Leominster, 1950.

ABOVE: Council, 1961. BELOW: The cattle judging rings at Hartpury in 1957.

ABOVE: Senior Officials of the Society in 1958: Glynne Hastings, Mr Millichap, Earl St Aldwyn, Emlyn Morgan, Joe Terry. BELOW: The Council of the Society in 1958.

LEFT: HM Queen Elizabeth the Queen Mother visited the show in 1958, escorted by the President, J. R. Hugh Sumner CBE. RIGHT: Glynne Hastings, Secretary 1946-1972. BELOW: Long Service Awards. RIGHT: The Queen Mother talks to Pat Smythe.

ABOVE: Awaiting the arrival of the Queen Mother: left to right — Jack Bellamy, Geoffrey Bright, W. H. Limbrick, Joe Terry and Glynne Hastings. BELOW: The opening ceremony 1964: a fanfare is sounded by the trumpeters of the Lancashire Fusiliers; Long Service Award winners are seated in the background.

ABOVE: The Showground 'Bus Service'. LEFT: Receiving the 1967 Long Service Award from the President, Lord Cobham. RIGHT: Vic and Joy Withers, exhibitors of Gloucester Old Spot pigs for many years.

A view of the showyard in 1967.

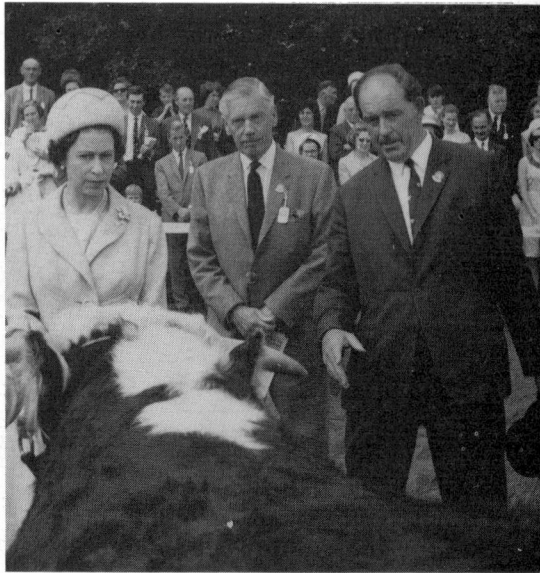

LEFT: Timothy Downes welcomes HM the Queen to the show in 1968. Others in the picture are (left to right) President Lord Rennel, L. Joseland, Sidney Andrews, Frank Smith and Glynne Hastings. RIGHT: Frank Smith explains the finer points of Hereford cattle to the Queen. BELOW: Her Majesty casts a knowledgeable eye over the Corgi class.

LEFT: 'Oh no, not another downpour!' RIGHT: HRH the Princess Alexandra is a popular visitor, who has attended the show three times. The conditions on her first visit in 1977 did not deter her. BELOW: The interior of the flower tent.

ABOVE: The scene that Show Secretaries dread and BELOW: the one they dream about.

BEHIND THE SCENES

No history of the Society would be complete without mention of some of the individuals and characters, many of them long forgotten, who have all played a part in bringing the Society to its present eminence.

A great debt is owed to such families as the Cobhams, the Cotterells, the Berkeleys, the Lechmeres and many others. Going through the records their names appear time after time as playing a leading part in their County and in the combined Society, ever since the earliest days.

Then there were yeoman farmers such as the Garnes, well-known throughout the Cotswolds as breeders of Cotswold sheep, the Hobbs family, famous for Shorthorn cattle and Oxford Down Sheep and in Hereford there were the Price, Taylor and Griffiths families, breeders of Hereford cattle. With pigs there were the Spencers, Rymans, Withers and Williams, all of whom exhibited for many years.

Also playng a prominent part were the clergy. Nowadays we forget how interwoven and what a prominent part the church played in the agricultural industry. In many parishes the parson farmed 'The Glebe' and knew as much as anyone about the soil. Being one of the few educated men in the parish he was looked to for guidance in matters temporal as well as spiritual. Thus we find Rev John Duncumb as one of the founders of the Herefordshire Society.

There was often mention of 'The Tithe' and suggestions that it be abolished, when times were hard. Indeed it must have been galling to see the parson getting fat on a good Glebe and then having to turn round and hand over a tenth of your produce annually.

While the names of the great men of stock breeding are recorded in the old catalogues, much of their effort would have been set at naught without the skill and devotion of their herdsmen. They used literally to live with their charges but alas, with most of them, their names have long since passed into obscurity.

Before road transport became readily available, stock used to travel to the show by rail. The larger animals travelled in cattle trucks and walked from the station to the showground. The smaller animals such as pigs and sheep were carried in crates and transported from the station to the showground on railway company drays.

Also travelling with the stock would be the stockmen and their 'Show Box'. The show box would be packed with necessities for the stock and stockman, such as brushes, curry combs, shampoo and the stockman's secret prescription for putting a bit of extra bloom on an animal. The name of the herd and its owner were painted on the lid. Inside the lid were pinned the herdsmen's rosettes, won during the season's show circuit. To this day the showbox is still a necessary part of the stockman's kit.

As soon as the animals were housed in their stalls and pens, they would have tarpaulin sheets, painted with the owner's name, hung around them. This was to hide them from their fellow competitors, until after the judging. Anyone attempting to look behind someone else's sheet was definitely discouraged, unless he was a close friend of the person concerned.

In this enclosure would go the show box and the herdsman himself, sleeping alongside the animals. In the case of pigs and sheep an extra pen would be booked for accommodation. All this proximity of herdsmen and animals was to discourage someone from nobbling a possible winner during the night. Some of the herdsmen were paid by results and were not scrupulous about how they obtained them.

Mostly there was, and still is a great bond between stockmen. Very often friendships could go a little bit over the top. It often happened that a female animal would come into season during the show and when this happened the servcies of a friendly bull or boar were often given, in exchange for a visit to the beer tent next day. It is most unlikely that the owners ever knew anything about these goings on; consequently many noble sires and dams maybe did not always carry the pedigree they were supposed to.

Few visitors to the show, when they see the immaculately turned out animals in the judging rings, can imagine the effort required to bring them to the peak of perfection. Preparations will have started many months beforehand. The breeding programme will have been worked out years in advance and the feeding will be regulated to bring them to the peak of condition on the day.

With sheep the shepherd will have been busy with a pair of clippers, trimming its fleece, so that the animals' good shape is accentuated.

With pigs, as with all animals, hours will be spent exercising so as to get the animal to walk and stand right. If this was not done the pigs especially would vanish out of sight, as soon as they were let out of the pen.

On the morning of the judging, the herdsmen will be up and about, feeding, grooming and exercising at an early hour. A feeling of tension builds up as the hour of judging approaches but, as soon as the judging starts, the spring is released. Everyone is too intent on getting their animal to stand right and get it in positon to catch the judge's eye to be affected by nerves.

After the judging is over there are handshakes, congratulations and celebrations in the bar. Then there is the cattle parade in the main arena, the animals having to be kept clean and in good order for the public to see.

At last it is the third day of the show, the final cattle parade and then it is pack up and go home and get ready for next year, or maybe go on to the next show.

These scenes have altered little over the years. Modern cattle lorries have done away with the hassle of rail travel and in addition they double up as sleeping quarters for the stockmen, although sleeping cubicles are available and the stockmen no longer have to bed down with their charges.

A regular show exhibitor in the immediate pre- and post-war years was the 'Cataline Cow'. This was an artificial cow, mounted on a lorry and was operated by a Mr Rowland Fursdon and his assistant, Freddie. Its purpose was to extole the virtues of a patent medicine called 'Cataline'.

Fursdon would stand declaiming what a wonderful miracle cure it was, saying something like this: 'The farmer went into the field one day and there was Daisy looking far from well. "Whatever is the matter Daisy" he would say'.

Freddie, who would be sitting at the control console, out of sight and watching through a peep-hole, pulled a few levers and Daisy would roll her eyes, put out her tongue and shake her head.

'Whatever's the matter', said Fursdon and Daily would jerk her head in the direction of her udder, and lift up her hind leg so that her mammary glands were plainly visible. In the meantime Freddie would be pumping away like mad on a pump to blow up Daisy's udder, which was inflatable. It would blow up to an alarming size and assume an angry red hue.

'Ah so that's the trouble', said Fursdon, 'Cataline will soon put that right'. He then went through the motions of pouring a bottle down Daisy's gullet, Freddie pulled the plug out, the udder resumed its proper shape and the demonstration was over.

Fursdon was a first-class showman and the demonstration always drew a large crowd, especially popular with children. Alas, some sacreligious stockman, possibly a dissatisfied customer, got round Daisy during the hours of darkness and stuck a pin in her udder.

They were all there waiting for the first demonstration next morning. When it came to the udder swelling you could hear Freddie cursing and groaning in the background and Fursdon whispering out of the corner of his mouth 'Pump harder, Freddie'.

After that there were no more demonstrations, which was a pity because, whatever the merits of Cataline, it was a main show attraction. They pleaded old age when asked to put Daisy through her paces.

In the weeks prior to the show the tenting contractors and permanent staff will have been busy. The grass must first be mown over the whole showground, temporary stands erected; cow stalls, sheep and pig pens, the grandstand, all have to ready in time for the opening.

During the week before the show, standholders will be coming in to set up their displays. This activity gradually builds up to the eve of the show and the organisers pray for fine weather. Rain will turn the temporary tracks and the ground adjoining the hard roads into a sea of mud. The Society is always adding to the permanent roadways but there never seem to be enough.

Anyone visiting the ground the evening before the show must wonder how everything can possible be ready on time. Huge lorries pass in a steady stream through the entrance gate to disgorge farm machinery, flowers, last minute arrivals of stand equipment and everything that goes to make up the show. Inevitably traffic jams occur and tempers evaporate, when someone finds that access to his stand is blocked.

The office staff behind the enquiry counter will be answering a whole barrage of questions and complaints. A water main has burst outside my stand — my next door neighbour is encroaching — I booked twenty feet and you have only given me fifteen — the electricity supply I ordered has not been connected — I haven't got enough entry passes for my staff — so it goes on long into the night.

Eventually and through sheer exhaustion, by around eleven o'clock everything quietens down. Those who are going home leave and those who are sleeping on the showground hope they can relax enough to get some sleep.

After all too short a night you awaken to a peacefulness that seemed to be impossible a short time ago. Horses are being cantered around the horse rings, to loosen up. Stands are looking neat and tidy, the chaos of last night has disappeared and the standholders, who the previous night had been stripped to the waist working on their stands, are now dressed in smart suits, perhaps watering a few plants or flicking imaginary specks of dust off gleaming machines.

Show stewards will be escorting their judges and other stewards will be ensuring that the stock is paraded in the judging ring, at the appropriate time.

As the day moves on the crowds pour in to do whatever they want to do. Some come to do business, some come to admire the stock and the arena events. Schoolchildren come on what we hope are educational trips, which seem to consist of collecting as many leaflets as possible. Others seem to do a lot of business in those tradestands that provide substantial refreshment.

Eventually the three days come to an end, the stock depart, the main arena events come to a close and the standholders start dismantling their stands and the flowers in the flower tent are sold off.

By next morning the showground is an untidy mess, stands half dismantled and half empty. Cattle and horse stalls are littered with mucky straw and the temporary shedding has a sad sagging look about it.

The clearing up operation goes on for several days. Repairs have to be made where necessary and if it has been a wet show, the main arena reinstated. The various committees hold post-mortems on what has happened and everybody gets ready for the next event to be held on the showground.

The Show must go on!

Mr Zhoa Ziyang led a Chinese delegation from Sichuan Province in 1979. He is here with the President, Sir Tatton Brinton. Ziyang became President of the People's Republic but was deposed after the Tiannemen Square Massacre in 1989.

LEFT: President Sir John Cotterell accepts the keys of the President's pavilion from the directors of National Westminster Bank. RIGHT: Ronnie Mills, horse steward and past Vice-President, presents the first prize in the sheep class. BELOW: Interior of the Sheep Building (Avon Hall).

ABOVE: Trotting races are a popular feature of the Show. LEFT: Mrs Shirley Sivell presents an award for the best turned out horse and cart. RIGHT: Mrs Hayes of Hayes Jewellers Ltd presents her firm's cup for Breeding Ponies. BELOW: The Society's facilities are regularly used for horse and pony sales.

ABOVE: The crosscut sawing competition. BELOW: Sheep dipping demonstration.

LEFT: The then Minister of Agriculture, Rt Hon Peter Walker MP, now Lord Walker, visits the bacon competition, accompanied by Geoffrey Ballard and Richard Collins, who obviously suspects something is 'going off'. RIGHT: Mr R. N. Seal, loyal supporter and critic of Society affairs, always aims to be first through the gate on the first day of the Show. BELOW: A pre-show trim, short back and sides! RIGHT: Sir John Cotterell presents the Cup for the champion Jersey, while Joe Beckett, Chief Dairy Steward, looks on.

LEFT: Norman Coward of the Midland Bank is perhaps setting an overdraft limit! RIGHT: The Horticultural Steward, Malcolm Hodges, (far right) watches Stephen Bailey receiving the Legal & General Assurance Cup. BELOW: A smart pony and trap turnout.

LEFT: HRH the Princess Royal with President Sir John and Lady Willison. RIGHT: The Princess Royal greets Sir Michael Higgs. BELOW: HRH the Prince of Wales with President Lord Vestey. RIGHT: The Prince of Wales is accompanied by Forestry Steward, Major Harvey Bathurst.

LEFT: HRH the Princess Margaret is introduced to Chief Executive, Lyn Downes by the President, Lady Holland-Martin. RIGHT: Princess Margaret presents the cup for the Champion Friesian, watched by Dairy Steward Ron Davies. CENTRE: Captain Mark Phillips presents long service awards. RIGHT: Long service award winners. BELOW: The Military Police display team.

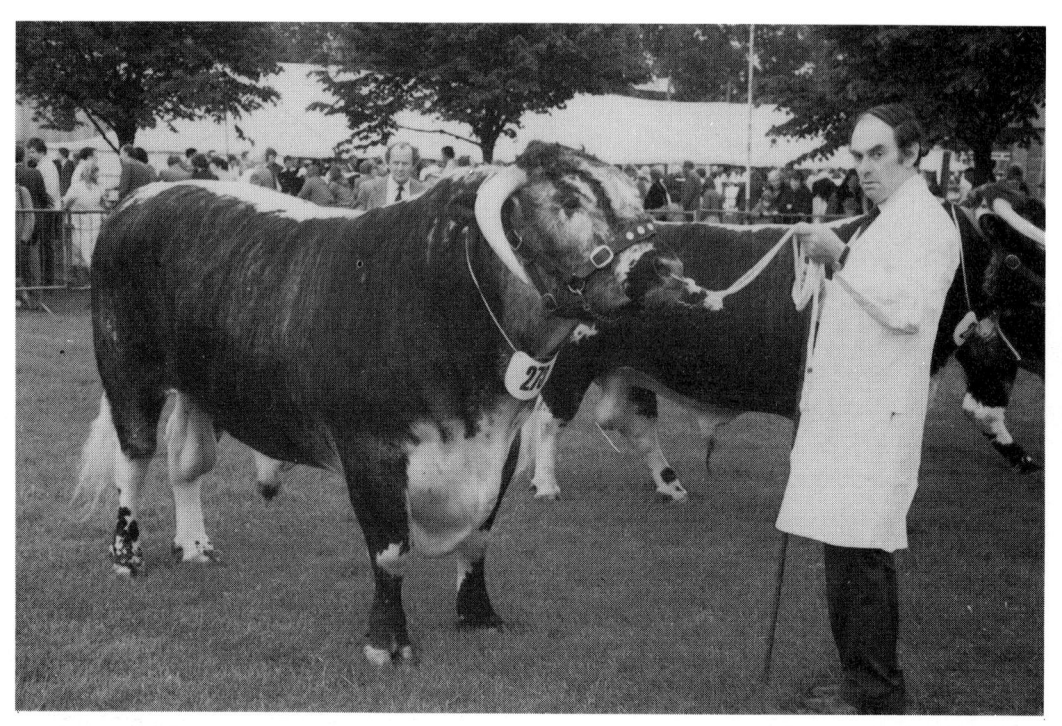

'And when did you last see your father?'

ABOVE: The Society actively encourages rare breeds, as witnessed by this magnificent Longhorn bull. BELOW: This cartoon appeared in the early years of the Charolais cattle importation, when a bull of doubtful pedigree was exhibited.

LEFT: A fine Rhode Island Red in the poultry section; the organiser of the Poultry Section is Mr G. B. Gee. RIGHT: The Chief Executive receives a sponsorship cheque from a representative of Messrs Paul's. Lyn Downes joined the Society staff in 1958; he was appointed MBE in 1992. BELOW: A satisfied exhibitor.

ABOVE: Parade of ponies in the main arena. BELOW: The West of England Ladies Kennel Club Show is a main event that takes place on the showground.

112

Endgame

Over the past two hundred years the emphasis and aims of the Societies have altered.

Until quite recently Mayors and local Council officials played a large and vital part in the organisation and wellbeing of the show, when it visited their city. In the early days the prosperity and trade of the cities was founded on a thriving market and agricultural industry. This is recognised in the agricultural symbols depicted in coats of arms up and down the country.

Local members of Parliament encouraged and took a large part in their local Societies. They looked upon them as a valuable opportunity to inform and be informed on current topics. In the early days the Societies spent a great deal of time discussing matters that affected the countryside. Today this function has been taken over by the National Farmers Union and the Country Landowners Association.

Exhibitors in the livestock classes have good cause to consider themselves hard done by, when they look at the value of the prize money they receive, compared to their counterparts of two hundred years ago.

When the Norfolk Parson James Woodforde died in 1803, three cows and a heifer that he owned were valued at twenty pounds the lot. Thus the award of a five pounds prize would have been something like the commercial value of the beast. Mr Thomas Moore of Cofton Hall won four first prizes at the first meeting of the Worcester Society, which would have been a terrific boost to his income. In an attempt to introduce an element of fairness the Hereford Society in 1829 brought in a rule that no exhibitors should be allowed to win a first prize two years running.

By 1850, as we have seen, the comercial value of a beast had risen to around £15. By this time the word 'Pedigree' was beginning to enter the cattle breeder's vocabulary and showing took on a different meaning. The relative cash value of the prize money became less and the return was achieved in the increased value in sales of breeding stock from prize winners.

This trend in livestock breeding sowed the seeds of its own destruction, albeit over a long period. By operating a closed herd book a point was reached where there were fewer bloodlines to choose from and further improvement impossible, without the introduction of fresh bloodlines from abroad or wherever.

If the trends that have overtaken pigs and poultry are anything to go by, the outlook is bleak. True we have a poultry show, but no-one would seriously consider running a large commercial flock of these old breeds.

Similarly with pigs. The breeding of this class of stock has largely been taken over by big breeding companies, whose sole aim is to turn out large numbers of hybrid animals to stock ever larger herds. True there are a few breeders to whom we are indebted for showing a few examples of our old traditional breeds. But whereas fifty years ago there were often eight classes to a breed and up to twenty pigs in a class, we are now lucky if we get twenty representativs of a breed.

Will the trend that has overtaken pig and poultry breeding be followed by cattle and sheep? The question should not be dismissed as unlikely. If this happens then, like the hybrid pig breeders, the hybrid cattle breeder will see little point in bringing animals to the show.

Certainly in the case of pigs and poultry, farmer interest in the show started to decline when the pedigree aspect of stock breeding gave way to hybridisation.

Following on from this the manufacturers of pig and poultry equipment no longer find it worth their while to put a big effort into exhibiting, other than a few odd bits of equipment on someone else's stand.

With the sheep the position is rather different; the production of sheep meat does not readily lend itself to intensive indoor production, as do other classes of livestock. Therefore, for the forseeable future, it would seem that our traditional native breeds may well hold their own, in the widely differing conditions of their traditional feeding grounds.

This leads on to the question, where does the future of the show lie if technological advances are going to overtake us, having in fact already done so in many instances? The commercial farmer has to be attracted to the show and feel that attendance is an essential scource of information for the success of his business. At the moment this is in doubt.

It could be that we have to some extent reached saturation point, as regards attracting farmer membership. Anyone who has tried recruiting members knows that it is far easier to sign up urban dwellers than it is farmers. This must be indicative of who is most attracted to the show. This opportunity of presenting a true picture of the agricultural industry, to the lay person, at a time when there is so much misinformation about, must be an important function of the Society. This will not happen unless true farmer interest is maintained.

It will have been noted how farming fortunes have ebbed and flowed over the past two hundred years. After each crisis or depression the tendency has been for the farm unit to enlarge and for there to be a great technological leap forward. There is no reason to suppose that this will be reversed.

The Society, in common with other show societies up and down the country, has reached a point where it wants someone or some society to be brave enough to break out of the familiar pattern. At the moment all societies are attempting too much in that they are trying to do a little bit for everyone and not very much for anyone.

We hope that people will always be attracted by the 'Country Fair' atmosphere and that the congenial surroundings will foster good business relationships between customer and supplier. This after all is what the show is all about, but it will not happen unless the show, or the Society, means something to the modern progressive farmer.

The phenomenal growth of the Spring Garden Show and the popularity of the National Sheep event, could well be pointers to the path that should be taken, with perhaps a number of specialist exhibitions held annually or bi-annually in conjunction with other Societies and organisations.

With new techniques developing in every branch of farming we could be on the verge of developments which will sweep away all that to which we have become accustomed. It is understandable that many of us have nostalgic longings for the past, but during the past two hundred years, changes so vast that they defy detailed enumeration, have passed over British agriculture. New practices, new implements, new methods are gradually adopted and so soon become familiar, that previous conditions are soon forgotten.

Britain, as an industrial nation, has seen many industries come and go, but agriculture will always be with us, though others may fail. This will always be so, because it serves the basic necessities of life, not only for human beings, but for the health of the environment in which we live.

This being so, Lord Lyttleton's speech, to the Worcestershire Agricultural Society in 1872, is even more important in today's political and environmental climate, one hundred and twenty-five years later, than it was when it was delivered at the time.

It should be engraved in letters of gold in every planning authority and Government department throughout the land and it is here repeated.

'A country like England, crowded with inhabitants and yet apparently not by any means arrived at the limit of its [agricultural] resources, in any respect, would be guilty of the most suicidal policy if at any time it failed to pay the greatest attention to the development of its own agricultural products. That a nation's capital is the health and proper use of its land'.

HRH the Princess Alexandra visited the show in 1989, and is here being presented to Mr David Burlingham by President Col U. Corbett.

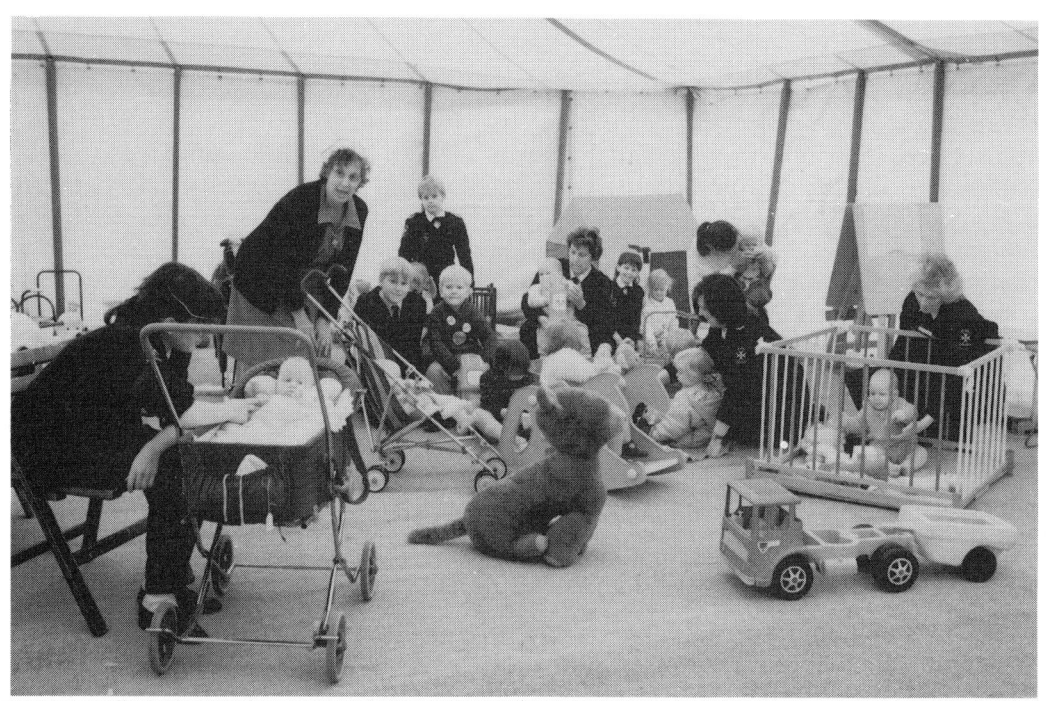

ABOVE: Children are also provided for. BELOW: Machinery lines, 1991.

LEFT: Mrs Phyllis Colwill, President 1988, receives the Severn Hall Mural from the donor, Neston Capper. RIGHT & BELOW: Vintage tractors and engines are a popular feature of the June Show.

ABOVE: A few of the Society's Trophies. BELOW: Cattle parade in the main arena.

ABOVE: The King's Troop Royal Horse Artillery were a popular attraction in 1989. BELOW: Members of Council, 1994.

Appendix A

Hereford Agricultural Society

Year	President	Where Show held	Cr/Dr	Balance £ s d
1798	Earl of Oxford	Broad Street		
1799	Earl of Oxford	Broad Street Three Shows		43 5 6
1800	Earl of Oxford	Broad Street Three Shows		35 16 6
1801	Earl of Oxford	Broad Street Three Shows		29 12 0
1802	Mr T. A. Knight	Broad Street Three Shows	Dr	3 6 6
1803	Mr T. A. Knight	Broad Street Three Shows	Dr	38 15 6
1804	Sir George Cornewell	Broad Street Three Shows		
		Leominster One Show	Dr	42 3 6
1805	Sir George Cornewell	Broad Street Two Shows	Dr	53 15 0
1806	Duke of Norfolk	Hereford 2, Leominster 1	Dr	1 7 6
1807	Mr R. P. Scudamore MP	Hereford 3, Leominster 1	Dr	29 9 0
1808	John Matthews	Hereford 3, Leominster 1		Unavailable
1809	T. P. Symonds MP	Hereford 3, Leominster 1		Unavailable
1809	T. P. Symonds MP	Hereford 3, Leominster 1		Unavailable
1810	Lt-Col Foley	Hereford 3, Leominster 1		Unavailable
1811-1813	Unavailable	Hereford		Unavailable
1814	Hon A. Foley MP	Hereford, Broad Street		Unavailable
1815	Edward Poole	Hereford, Broad Street		Unavailable
1816	Ben Biddulph	Hereford, Broad Street		Unavailable
1817	Unavailable	Hereford		Unavailable
1818	Mr Price			
1819-1822	Unavailable			
1823	Sit Anthony Lechmere			
1825	Unavailable	Hereford and Worcester joint show		
1826	Unavailable	Hereford	Dr	114 18 2
1827	Unavailable			
1828	Unavailable. Liabilities so large that Society wound up. New Hereford Society formed.			
1829	Sir J. Cotterell	Hereford, Broad Street		20 15 2
1830	James Phillips	Hereford, Broad Street		9 10 0
1831	Sir Robert Price	Hereford, Broad Street		5 5 0
1832	Kedgwin Hoskins MP	Broad Street, St Owen Street		17 12 8
1833	T. A. Knight	Broad Street, St Owen Street	Dr	8 8 0
1834	E. T. Foley	Broad Street, St Owen Street	Dr	18 18 0
1835	Sir G. Cornewell	Broad Street, St Owen Street	Dr	33 16 6
1836	Rev J. L. Penvyre	Mr Bosley's field		32 17 8
1837	Philip Jones	Mr Bulmer's field	Dr	13 2
1838	Philip Jones	Mr Bulmer's field	Dr	5 6 7
1839	Sir E. F. S. Stanhope	Barling's Yard, King Street		19 15 10
1840	Col Scudamore	Mr Bulmer's field		22 9 3
1841	Sir H. Hoskins	Mr Bulmer's field		1 14 3
1842	Earl Somers	Hereford		29 19 3
1843	Joseph Bailey MP	Hereford	Dr	32 6 7
1844	Richard Hereford	Hereford		40 7 2
1845	T. B. M. Baskerville MP	Hereford		5 1 10
1846	J. King	Hereford		33 7 10
1847	Chandos W. Hoskins	Hereford	Dr	7 6
1848	T. R. Haggitt MP	Hereford	Dr	7 19 2
1849	G. C. Lewis MP	Hereford		2 4
1850	H. M. Clifford MP	Hereford		29 17 9

Year	Name	Location		£	s	d
1851	Lord Bateman	Hereford	Dr	24	12	10
1852	T. W. Booker MP	Hereford		5	13	2
1853	C. S. Hanbury Bateman	Hereford	Dr	37	2	0
1854	H. Lee Warner	Hereford Market	Dr	66	0	2
1855	W. T. Keville Davies	Hereford Market	Dr	27	5	5
1856	Elias Chadwick	Hereford Market		57	19	1
1857	Rev Archer Clive	Hereford Market	Dr	36	9	7
1858	Robert Biddulph	Hereford Market	Dr	22	2	7
1859	C. W. Allen	Hereford Market	Dr	20	6	10
1860	George Clive MP	Hereford Market		12	0	7
1861	W. Money Kyrle	Hereford Market		44	19	6
1862	Sir George Velters Cornewell	Hereford Market	Dr	20	9	5
1863	J. H. Arkwright	Hereford Market			17	10
1864	H. Mildmay	Hereford Market		3	17	8
1865	Sir Josepeh Bailey	Hereford Market (No prizes)		136	7	0
1866	Sir Joseph Bailey	No show owing to rinderpest				
1867	Sir Joseph Bailey	Reduced Show	Dr	19	19	11
1868	Sir Joseph Bailey	Hereford Market	Dr	8	15	10
1869	Michael Biddulph MP	Hereford Market	Dr	55	3	10
1870	Sir Herbert Croft MP	Hereford Market		1	4	1
1871	Meysey Clive Esq	Mr Partington's meadow		40	4	4
1872	D. Peploe Peploe	Hereford Market	Dr	92	5	4
1873	Rev Sir J. H. Cornewell	Hereford Market	Dr	28	12	0
1874	A. R. Boughton Knight	Hereford Market and Mr Partington's meadow	Dr	323	7	7

Collected from Guarantors £322 17s 6d plus £1 for non-exhibition of stock, still left a debtor balance on the year of £12 2s 1d.

Year	Name	Location	£	s	d
1875	Richard Herefored	Hereford Market and meadow. Guarantors again had to pay £175 17s 11d to balance.			
1876	Lord Bateman	Hereford Market and meadow	44	5	1
1877	Joseph Pulley	Hereford Market and meadow	7	1	8
1878	Joseph Pulley	Kington	90	9	6
1879	James Rankin	Hereford	47	5	3
1880	Sir Henry Stanhope	Hay on Wye	120	11	3
1881	E. Pateshall	Leominster. No financial records available from now on.			
1882	Major Meysey Clive	Hereford			
1883	Crawshaw Bailey	Abergavenny			
1884	Stephen Robinson	Ledbury			
1995	J. H. Arkwright	Monmouth			
1886	Lord Tredegar	Brecon			
1887	Sir Joseph Bailey	Ross on Wye			
1888	Earl of Coventry	Tenbury			
1889	Lord Chesterfield	Hereford			
1890	J. W. Hastings	Malvern			
1891	Richard Green	Kington			
1892	Rev. G. H. Davenport	Pontypool			
1983	Earl of Coventry	Kidderminster			
1894	Joseph Pulley	Hereford			

Appendix B

Herefordshire & Worcestershire Agricultural Society

Year	President	Venue	Year	President	Venue
1895	Earl Beauchamp	Worcester	1909	H. J. Bailey Esq	Leominster
1896	Lord Llangattock	Monmouth	1910	Stanley Baldwin Esq MP	Worcester
1897	Lord Windsor	Redditch	1911	Lord Biddulph	Ledbury
1898	Sir Henry Lambert	Stourbridge	1212	Edward Partington Esq	Droitwich
1899	James Rankin Esq MP	Hereford	1913	C. T. Bailey Esq	Hereford
1900	H. J. Bailey Esq	Leominster	1914	Sir H. F. Grey	Worcester
1901	Earl of Coventry	Evesham	1915	Sir H. F. Grey	Worcester
1902	Sir J. R. G. Cotterell	Hereford	1916	Earl of Coventry	Hereford 2 days
1903	Earl Beauchamp	Worcester			
1904	Captain P. A. CLive MP	Ross on Wye	1917	Earl of Coventry	No Show
1905	Sir H. F. Lambert	Malvern	1918	Earl of Coventry	No Show
1906	Lord Windsor	Bromgsgrove	1919	Sir John Cotterell Bart	Hereford
1907	Earl of Chesterfield	Hereford	1920	Earl of Coventry	Worcester
1908	Lord Hindlip	Kidderminster	1921	Lord Cawley	Hereford

Appendix C

Three Counties Agricultural Society

Year	President	Venue	Year	President	Venue
1922	Colonel J. Russell-Kerr	Gloucester	1965	Colonel J. F. Maclean	Hereford
1923	Earl of Coventry	Malvern	1966	Miss O. Lloyd Baker CBE	Gloucester
1924	Stanley Baldwin MP	Worcester	1967	Viscount Cobham KG, GCMG	Worcester
1925	Lord Somers	Hereford	1968	Brig A. F. L. Clive DSO, MC, DL, JP	Hereford
1926	The Duke of Beaufort	Gloucester			
1927	Lord Doverdale	Worcester	1969	The Duke fo Beaufort KG, PC GCVO	Gloucester
1928	Sir John Cotterell Bt	Hereford			
1929	HRH The Duke of Gloucester	Gloucester	1970	Sir Michael Higgs DL	Worcester
1930	Viscount Cobham	Worcester	1971	The Lord Bishop of Hereford	Hereford
1931	Viscount Hereford	Hereford	1972	Mr R. J. Berkeley, TD, MFH	Gloucester
1832	Earl Bathurst	Gloucester	1973	Lord Sandys DL	Worcester
1933	Major W. Harcourt Webb	Worcester	1974	Lt Cmdr J. H. S. Lucas-Scudamore	Hereford
1934	Lord Somers	Hereford			
1935	Sir Percival Marling VC	Gloucester	1975	Lord Dulverton CBE, TD	Gloucester
1936	Earl of Dudley	Worcester	1976	Sir John Willison OBE, DL	Worcester
1937	Lt-Co R. N. H. Verdin	Hereford	1977	Capt T. R. Dunne HML	Hereford
1938	The Viscount Bledisloe	Gloucester	1978	Lord Vestey	Gloucester
1939	Captain R. G. W. Berkley	Worcester	1979	Sir Tatton Brinton	Worcester
1940	Lord Somers	Hereford	1980	Sir John Cotterell	Hereford
In abeyance during hostilities.			1981	HRH The Princess Anne	Gloucester
1947	Sir R. C. G. Cotterell Bt	Hereford	1982	Sir Berwick Lechmere	Worcester
1948	The Viscount Bledisloe	Gloucester	1983	Mr S. C. Andrews MBE	Hereford
1949	Captain R. G. W. Berkeley	Worcester	1984	Earl of Gainsborough	Gloucester
1950	Sir R. C. G. Cotterell Bt	Hereford	1985	Lady R. Holland-Martin	Worcester
1951	His Grace the Duke of Beaufort	Gloucester	1986	Mr Bertram Bulmer	Hereford
			1987	Colonel M. St J. V. Gibbs	Gloucester
1952	Admiral Sir William Tennant	Worcester	1988	Mrs P. M. Colwill	Worcester
1953	Lord Rennel of Rodd	Hereford	1989	Lt Col U. Corbett C.B.E, DSO	Hereford
1954	Earl of St Aldwyn	Gloucester	1990	Mr Oscar H. Colburn, CBE, JP DL	Gloucester
1955	Sir Chad Woodward	Worcester			

1956	Finola, Lady Somers CBE	Hereford	1991	Viscountess Cobham DL		Worcester
1957	Lt Col John Godman CBE	Gloucester	1992	Lady Cotterell		Hereford
1958	J. R. Sumner CBE	Worcester	1993	Her Grace The Duchess of Beaufort		Gloucester
1959	Mrs A. Simmons MHF	Hereford				
1960	Lord Trevethin & Oaksey	Gloucester	1994	Lord Walker of Worcester MBE, PC		Worcester
1961	The Lord Bishop of Worcester	Worcester	1994	Major B. A. F. Hervey-Bathurst		Hereford
1962	Sir Archer Baldwin MC	Hereford				
1963	Lord Banbury of Southam	Gloucester	1966	Mr H. W. G. Elwes HML		Gloucester
1964	The Earl Beauchamp	Worcester	1977	Lord Plumb DL, MEP		Worcester

Note: 1958 was the first year that the Show was held on the permanent site at Malvern. Hereafter the Counties mentioned acted as host county for the year.

Appendix D

Venues and entries where known, of the Gloucester Agricultural Society

Year	Venue / Note					
1829	Cirencester.					
1830-33	Cirencester Agricultural Society only.					
1834	First show of the Gloucestershire Agricultural Society, Gloucester Market.					
1835-52	Shows held in Gloucester Market, also occasional ploughing matches.					
1853	Combined with Royal Show at Gloucester					
1854-61	Combined with Circencester Society and held shows alternately at Cirencester and Gloucester.					
1862	First show of 'Enlarged County of Gloucester Society, held in 'The People's Park Gloucester'. 19 classes of cattle, 20 classes of sheep, 4 pigs, 5 horses.					
1863	Cirencester.					
1864	Cheltenham. 50 entries of cattle, sheep, pigs, horses, implements and a dog show.					
1865	Tewkesbury in conjunction with the Annual Regatta.					
1866	Show cancelled because of rinderpest.					
1867	Cheltenham in conjunction with the Cheltenham Horticultural Society.					
1868	Gloucester					
1869	No Show					
1870	Cirencester					
1871	Cheltenham					

Year	Venue	Cattle	Sheep	Pigs	Horses	Total
1872	Gloucester	84	56	44	76	260
1873	Bristol	97	68	94	124	383
1874	Tewkesbury	84	55	74	157	370
1875	Cirencester	89	43	73	139	343
1876	Cheltenham	77	29	71	99	276
1877	Dursley	55	38	53	99	245
1878	No Show					
1879	Cheltenham	41	36	51	156	284
1880	Gloucester	73	24	48	122	267
1881	Cirencester	105	41	55	110	311
1882	Cheltenham	148	33	71	154	406
1883	Berkeley	152	50	45	75	322
1884	Stroud	128	51	40	116	335
1885	Gloucester	104	37	62	89	292
1886	Cirencester	98	47	60	115	320
1887	Cheltenham	134	72	45	119	370
1888	Moreton	135	60	51	160	406
1889	Bristol	191	52	44	134	421
1890	No Show					
1891	No Show					

1892	Stroud	113	46	32	124	315
1893	Gloucester in combination with the Bath and West.					
1894	Cheltenham	116	77	S. Fever	144	337
1895	Cirencester	140	65	S. Fever	148	353
1896	No Show					
1897	Gloucester	147	43	55	158	403
1898	Cheltenham	152	58	S. Fever	200	404
1899	Cirencester	120	55	35	183	392
1900	Gloucester	141	52	S. Fever	151	344
1901	Cheltenham	155	58	S. Fever	151	364
1902	Tewkesbury	235	70	59	141	505
1903	Cirencester	156	60	S. Fever	144	360
1904	Gloucester	199	56	29	130	414
1905	Cheltenham	169	67	63	128	427
1906	Combined with Bath and West Society at Swindon.					
1907	Stroud	175	52	S. Fever	105	332
1908	Cirencester	183	74	61	100	418
1909	Combined with Royal Show at Gloucester.					
1910	Cheltenham	172	57	39	126	388
1911	Gloucester	160	40	23	134	347
1912	Stroud	154	42	23	113	332
1913	No Show					
1914	Cheltenham	141	53	S. Fever	123	317

Besides the above entries in the farm animal classes there were on many occasions classes for toy and sheep dogs, butter making, cheese making, flower shows, implement classes, jumping and driving and many others.

Unfortunately there are no reliable attendance records.

Appendix E

Venues of Worcester Agricultural Society

1816-1826	Two shows held annually, In Mr Hope's field 'At the top of the Tything adjoining the Shrubbery'.
1827-1829	Society Defunct.
1830	Attempt to revive the Society was unsuccessful.
1831-1837	Society Defunct.
1838	Ploughing match at Hallow.
1839-1862	Shows held in Worcester.
1839 & 1840	Ploughing matches at Hallow.
1863	Amalgamated with the Royal Show at Middle Battenhall.
1864	Astwood
1865	Cancelled because of rinderpest.
1866	Cancelled because of rinderpest.
1867	Worcester
1868	Worcester
1869	Henwick
1870	Kidderminster
1871	Malvern
1872	Stourbridge
1873	Evesham
1874	Dudley
1875	Worcester
1876	No Show because of competition from Royal at Birmingham and Bath and West at Hereford.
1877	Kidderminster

1878	Bromsgrove
1879	Malvern
1880	Combined with Bath and West at Worcester.
1881	Stourbridge
1882	Dudley
1883	Worcester
1884	Pershore
1885	Redditch
1886-1893	The Society ceased to exist.
1894	Revived and amalgamated with Hereford Agricultural Society.

Appendix F

Secretaries of the Herefordshire Agricultural Society, the Hereford and Worcester Agricultural Society and the Three Counties Agricultural Society.

Hereford Agricultural Society:
1797-1835	Rev J. Duncumb
1835–1836	Mr Bell
1836-1852	James Fowler
1852-1874	James Owen Thomas Fowler
1874-1881	Thomas Duckham
1881-1902	Alfred Edwards, also Secretary to the Hereford and Worcester Agricultural Society on amalgamation.
1898-1902	T. H. Edwards, Assistant Secretary.
1902-1946	T. H. Edwards, Secretary to the Three Counties Agricultural Society.
1946-1972	Glynne Hastings
1972-	L. M. Downes MBE. Secretary, later Chief Executive, Three Counties Agricultural Society.

Secretaries of the Worcestershire Agricultural Society:
1816-1819	Jeremiah Herbert
1819-1826	George Bentley
1826-1838	No Shows, Society Defunct.
1838-1843	John Crane Nott
1843-?	Edward Larkin
?-1869	Edward T. Goldingham
1869-1885	Mr Buck

Secretaries of the Gloucestershire Agricultural Society, (unfortunately their dates of appointment are not known):

John Kemp, A. G. Jones, Henry J. Cochrane, Edward Trinder, Robert Anderseon and L. W. Wrighley. The last three are known to have been agents to the Bathurst estate.

Appendix G

Statement of Cash Takings and Attendance

	Year	Gate Takings £	Grandstand Takings £	Car Park Takings £	Trade Stand Fees £	Attenance	Venue	Gain £	Loss £
	1922	3,965	535	83	—	27,373	Gloucester	1,905	—
	1923	4,031	534	17	—	26,637	Malvern	1,234	—
a.	1924	—	—	—	—	—	Worcester	—	2,144
	1925	3,667	543	146	—	24,513	Hereford	293	—
	1826	3,675	689	237	—	35,473	Gloucester	988	—
	1927	3,327	686	151	—	25,163	Worcester	555	—
	1928	3,134	520	202	—	24,831	Hereford	—	271
	1929	3,407	504	217	—	23,114	Gloucester	501	—
	1930	3,774	723	203	—	28,517	Worcester	659	—
	1931	2,964	517	170	—	22,582	Hereford	—	295
	1932	3,213	607	325	—	24,649	Gloucester	399	—
	1933	2,735	553	241	—	20,799	Worcester	56	—
	1934	3,875	682	378	—	29,989	Hereford	1,337	—
	1935	4,156	785	488	—	32,238	Gloucester	1,407	—
	1936	3,361	667	391	—	28,678	Worcester	899	—
	1937	3,399	600	320	—	29,592	Hereford	265	—
	1938	4,095	708	585	—	31,320	Gloucester	1,390	—
	1939	3,002	441	385	—	22,891	Worcester	267	—
	1940-1946 No Show was held during the War Years								
b.	1947	7,664	953	1,533	2,939	39,433	Hereford	2,853	—
	1948	16,215	2,378	2,598	6,293	63,422	Gloucester	9,675	—
	1949	12,148	1,726	2,420	8,526	47,570	Worcester	3,801	—
	1950	11,096	1,518	1,600	9,974	41,968	Leominster	—	2,753
	1951	12,951	1,310	2,390	12,102	47,847	Gloucester	2,951	—
c.	1952	9,861	1,037	2,173	11,975	39,786	Worcester	779	—
	1953	11,526	1,288	2,058	12,319	46,947	Hereford	3,279	—
	1954	11,821	1,243	2,489	13,393	40,683	Gloucester	417	—
	1955	11,270	1,152	2,773	14,604	45,782	Worcester	625	—
	1956	12,855	992	2,599	16,174	54,170	Hereford	1,385	—
d.	1957	11,548	953	2,258	13,153	31,705	Gloucester	—	3,176
e.	1958	15,800	1,188	2,295	17,425	48,152	Malvern(W)	5,320	—
	1959	17,106	1,365	3,252	17,626	55,436	Malvern(H)	4,223	—
	1960	17,259	1,866	3,659	18,037	55,713	Malvern(G)	1,690	—
	1961	17,502	1,512	3,815	18,368	57,314	Malvern(W)	6,490	—
	1962	22,030	1,771	4,412	17,950	61,210	Malvern(H)	10,509	—
	1963	16,086	1,297	3,975	18,616	50,469	Malvern(G)	2,027	—
	1964	17,201	1,001	4,243	18,343	57,763	Malvern(W)	7,781	—
	1965	15,646	881	4,161	18,112	50,693	Malvern(H)	1,389	—
	1966	16,710	1,012	4,389	17,543	53,033	Malvern(G)	3,175	—
	1967	16,438	952	4,263	16,388	52,586	Malvern(W)	3,330	—
	1968	18,256	1,012	4,442	16,121	55,378	Malvern(H)	2,064	—

	Year								
	1969	21,854	1,249	4,358	17,076	55,754	Malvern(G)	3,685	—
	1970	19,995	795	4,129	16,429	51,640	Malvern(W)	—	495
	1971	19,957	851	4,563	17,469	48,512	Malvern(H)	—	924
	1972	27,007	1,019	7,178	18,272	53,911	Malvern(G)	10,254	—
	1973	31,463	1,603	8,969	20,166	62,141	Malvern(W)	15,242	—
	1974	37,094	1,491	10,371	24,178	68,386	Malvern(H)	20,194	—
	1975	44,788	1,692	9,311	28,549	67,208	Malvern(G)	18,285	—
	1976	52,659	1,968	12,092	34,204	65,835	Malvern(W)	17,934	—
	1977	44,865	1,396	9,470	39,868	47,341	Malvern(H)	5,763	—
	1978	84,791	2,527	19,007	45,615	72,271	Malvern(G)	54,225	—
	1979	96,385	2,276	707	59,631	63,865	Malvern(W)	45,511	—
	1980	117,013	2,715	—	71,625	68,095	Malvern(H)	54,409	—
	1981	135,000	2,923	—	84,846	66,463	Malvern(G)	45,362	—
	1982	139,000	2,900	—	99,100	69,537	Malvern(W)	69,793	—
f.	1983	133,000	2,300	—	103,600	103,286	Malvern(H)	54,264	—
	1984	143,000	2,100	—	111,800	106,682	Malvern(G)	35,757	—
	1985	157,116	1,800	—	125,492	106,708	Malvern(W)	34,650	—
	1986	148,460	2,196	—	129,478	106,594	Malvern(H)	10,487	—
	1987	176,854	1,983	—	141,744	107,700	Malvern(G)	64,075	—
	1988	178,796	2,136	—	149,331	109,035	Malvern(W)	35,827	—
	1989	216,127	1,948	—	167,776	112,262	Malvern(H)	17,242	—
	1990	230,986	1,862	—	177,840	111,040	Malvern(G)	1,912	—
	1991	200,969	1,557	—	198,618	104,587	Malvern(W)	419	—
	1992	217,546	1,675	—	211,377	108,025	Malvern(H)	35,125	—
	1993	180,109	1,389	—	216,252	97,681	Malvern(G)	13,556	—
	1994	228,109	1,654	—	221,280	106,193	Malvern(W)	77,366	—
	1995	236,281	937	—	239,700	104,977	Malvern(H) 1	25,603	—

Notes (a) Show abandoned owing to floods.

(b) Show postponed one month owing to foot and mouth disease.

(c) No cloven hoofed classes owing to foot and mouth disease.

(d) Only two-day Show owing to petrol restrictions because of Suez Crisis.

(e) First Show on permanent Site at Malvern.

(f) Prior to 1983, attendance figures shown are paid attendance only. From 1983 total attendance is shown.

INDEX

All figures in *italics* refer to illustrations.

Abergavenny 25
Admiralty, Lords of 17
Agriculture Engineers Assoc ..74
Alcester 4
Alexandra, HRH Princess *96, 97,115*
Allen, paper on threshing 48
Allhambra Company 37
Allies, George 35
Allsbrook, A. 74
Allsop, Mr 38
Alton Beauchamp 34
Amblecote Church 39
Ampnett 53
Anderson, R. 69
Andrews, Sidney 96
Angel Street, Worcester 37
Apperley, W.H. 22
Arch, Joseph 39
Aumale, Duc d' 38
Avon Hall *103*
Aylesbury Dairy Co 40
Bailey, Stephen *107*
Baker, J.T. Lloyd 54
    T.B. Lloyd 58
Baldwin, A. 74
Ballard, Geoffrey *106*
Ballingham Hall 37
Bamford's 45
Bank of England 9
Barneby, Sheriff 21
Bateman, Lord 25
Bath & West Show ..17,23,39,40 45,62,73,74
    Society 46
Bathurst, Earl 53,57,60,63
    Hon 58
    Maj Harvey *108*
Beauchamp, Earl 36,38,44
Beaufort, Duke of 54,56
Beckett, Joe *106*
Bellamy, J. 74.*93*
Bennett, Mr 53
Bentley, George 34
Berkeley 61,99
    Hon C.F. 58
    Society 58
Berkley, Grantley 54,57
Berkshire (pigs) 45,59
Berrington House 83,85,87
Berrows Journal 32
Berry, Rev Henry 34
Bicentennial panel *2*
Biddulph, Ben 15,18
Blackmore Park Road 86
Bledisloe, Viscount 83,*87*
Bleriot 45
Board of Agriculture 46
Bosley, Mr 20,*31*
Bosley's Field 20
Boughton Park 72
Bowley, Mr 57

Boxsidge, Mr 22
Bradstock, Thomas 22
    P.E. 74
Bray, G.H. 74
Brazier, John 34
Brecon 25,45
Bredon Society 58
Briar, Wyndham 26
Bright, Geoffrey *93*
Bristol 57
Brinton, Sir Tatton *102*
Brittain, C. 22
Britten, Mr 43
    W.G.C. 74
Broad, Benjamin 17,18
    Rat Trap *28*
    Street, Hereford 9,15,19,20
Brocklehurst, H. Dent- 69
Bromsgrove 40
    Farmers' Club 41
Brown, James 55
Bruton, G.N. 84
    Sir James 69
Buck, Mr 24
Buckley, John 20
Bulmer, Bertram 27
    Messrs 20
Bulmer's Meadow 20,*31*
Burlingham, David 115
    H. & Co 38,*81*
Burlton, P. 22
Byster's Gate, Hereford 19
Cadle, Clement 37
Caddleworth 53
Caledonia Bank 11
Cally, Anne 18
Cambridge Roll 18
Canon Frome 45
Capper, Neston 117
Caravan & Camping Club 86
Carpenter, Mr 22
Carrington, Lord 16
Cartoon *110*
Castleton Farm, Clifford 17
Cataline Cow 100
Cattle Diseases Prevention
    Act 23
    Parade *77,118*
    Judging 90
Central Chamber of
    Agriculture 44
Chaddesley Corbett Farmers'
    Club 41
Charlton Park 62
Charolais *110*
Cheese, High Sheriff 22
Cheltenham Chronicle 71
Cheltenham Show 60,62,63, 72,86
    Horticultural Soc 60
Chesterfield, Lord 50
Chew, A.H. 74
children *116*
Churchdown 55

Church's Garden, Mr 56
Cirencester 53,59
    Agricultural Soc 53,56,57
    Assoc 54
Clayton, Captain 45
Clifford, H.C. 54
    Lt Col 22
Clifton on Teme Agricultural
    Soc. 41
City Rifle Co 58
Cobham 99
    Lord (1) *13*
    (2) *14*,45,*94*
Cochrane, Mr 60
Codrington, Sir C.W. 54,57
Cofton Hall 33,113
Collins, Richard *116*
Colwill, Mrs Phyllis *117*
Commercial Road, Hereford ..19
Cooke, Thomas 19
Coney, W. 37
Corbett, Col U.H. 115
Corn Exchange, Leominster ..83
    Worcester 37
    Laws, Repeal of ...10,19,21,36
Cornewell, ...15
    Sir George 17
Cornwall, Rev Sir George ..24,25
communications 86
Cotswold Ploughing Match 42
    sheep 36,59,99
    ram 67
Cotterell, ...15,99
    Sir G. 17
    J. 19
    John (1) *13*
    (2) *13,103,106*
Council 71,84,*90,91*,119
Coward, Norman 88
Cridlan, Mr 63
Country Landowners'
    Assoc 113
Coventry, Earl of 36,38
Creed, Mr 57
Crimean War 10
Cripps, Joseph 53,57
Croft Castle 22
Cross, Kington 22
crosscut saw *105*
Crowie 35
Crown Hotel, Worcester ...34,35
Curtler, Mr 36
Cuss, Mr 69
Dairy School 26
'Daisy' *101*
Davies, John 22
    Mr 22
    Rev Canon 60
    Ron *109*
Davis, J. (medal) 28
Davenport, Rev 43
Daw, Mr 22
Deighton, Mr 35
Devereux Wootton 20

Devon Cattle 11,58,159
'Diabolo' 45
Dowdeswell, Mr 38
    W.E. 60
Downes, L.M. 85,*109,111*
    Tim *96*
Dog Show *81*
dinner, cartoon *14*
Ducie, Earl of 56
Duckham, Thomas *12,23,24*
Dudley 39,40
    Earl of 38
Dumbleton 57
Duncumb, Rev John 15,99
Durham Ox 16
Dursley 61
Eardisland 22
Earles Croome 33
early implements *8*
Edmonds, Mr 53
Edward VII 62
Edwards, Alfred 25,44
    H.M. 25
    G.H. 74
    T.H. 25,44,74,78,83
Elmley, Viscount 33
Elmstone Hardwicke *9,14,*85
Elizabeth, HM Queen
    Mother 85,*92*
    II, HM Queen 85,*96*
Elwes, Mr 57
Emlyn, Lord 57
Enclosure *9,14*
Engall, Mr 60
Ensilage Press Co 41
Essex, G.R. 40
    Show 39
Evans, David 85,86
    William 55
Evesham 33,46
    Society 37,39
Eyton, H. 22,23
Fairford 56
farm operations *8*
Feckenham Lodge 33
Ferriers, Baron de 60
Firs Farm, Malvern 85
Fitzhardinge, Lord 57
Fleece Inn, Gloucester 54
Foley, Hon A. 18
    Hon E. 33
    Lady 22
    Lord 33
Foot and Mouth 24,40
Fowler's Steam Plough *42*
Foxwell, Mr 37
floods 72,76
Flower Tent *97*
Fraser, James 39
Freddie *100,101*
Free Traders 21
French Revolution 9
Frocester Court 56
Fursdon, Rowland *100,*10

Garne 99	assets *49,50*	Kidderminster...................38,40	Supply 83
W. *63,67*	Bromsgrove 1906............*51*	King Street, Hereford............20	Mitford, A.B., Freeman 61
Gee, G.B............................*111*	Declaration of Trust........*50*	Kings Acre, Hereford.............21	Moore, Thomas 33,113
George IV 60	dignitaries *51*	Square, Gloucester.............56	Monkhouse, Mr 20
Girdlestone, Canon..............39	Malvern 1914...................*50*	King's Arms, Leominster........17	Monmouth 25,45
Glamorgan 25	Officers & Council............*50*	Kingscote, Colonel................58	Mop Fairs 36
Glasgow Bank 11	opening ceremony	Kington 25	Moreton, Hon Henry.............53
Glebe, the............................99	1907 ....*50*	Kite, S...................................*69*	on Lugg...............................22
Gloucester, Bishop of...........39	Show 1908.......................*47*	Knight, Mr.............................18	in Marsh.............................61
College of Agriculture 85	Hereford bull........................29	Knightwick Farm 40	Society................................58
Duke of 72,73,78	cattle..................10,38,40,44,	& District Farmers' Club....41	Morgan, Emlyn *91*
Wagon Co 40	45,58,59,99	*Labourers' Union Chronicle*.......39	Napoleon 9
Society ....10 *passim*	City Council 23,24	Lamb Inn, Cheltenham 54	Napoleonic Wars..............9,15
classes..............................64	Corn Market 19	Lancashire Fusiliers *94*	National Farmers' Debt........19
judges...............................68	County Council 26	Langdale Wood (House)........86	Union...............................113
membership list *52*	Herd Book 20,23,25	Larkin, Mr 36	Sheep Assoc 86
minutes *65*	& Breed Soc 22	Larkworthy, J. & Co...............38	event................................114
programme, 1911 ....*68*	Lord 17	Lawrence, John.....................55	Nelson, Admiral Lord............60
Show 1865.......................*65*	map *31*	Lechmere *99*	Newcombe, Mr....................57
1883..............................*68*	ox 16	A...*33*	Newham Court.....................35
plan *66*	*Hereford Times*..........20,26,45,46	Sir Anthony *12,17,19*	New Cattle Market, Worcester
Gloucester Chronicle 54	Hewer, Mr.............................53	Berwick *12,19*	(letter) *28*,35
Journal 53,58	Hewitt, J.W...........................74	E. 36	(map).................................*31*
Gold Rush 10	Hicks, Sir William 54	Ledbury Society 24,25	Herefordshire Society 19,
Great Witley & District	Hicks-Beach, M.H.................63	Leen 22	*et seq*
Farmers' Club 41	Higgs, Sir Michael *108*	Legal & General Cup...........*107*	Market, Gloucester 56
Green Dragon Hotel,	High Town, Hereford............21	Leicester Sheep 38	Niblett, D.J............................56
Hereford 17,22,43	Highland Show 39	Leigh Court 40	Nicholson..............................45
Grenadier Guards.................58	Hindlip Hall 38	Leominster....16,18,25,43,84,*89*	Nigerian Farmers *79*
Greyhound Hotel, Hereford..17	Hoare, Sir R.C. 54	Society 24	Nile, Battle of.........................9
Griffiths..................................99	Hobbs....................................99	*Leominster Guide* 16	Nore, the 9
R. 20	R.W. 63,67,69	Lickhill Manor, Stourport 35	North Nibley.........................54
Guildhall, Worcester.............33	Hodges, Malcolm................*107*	Limbrick, W.H. *93*	Northleach Soc 58
Assembly Rooms 37	Mr 69	Lincoln (sheep)......................11	Nott, John Crane (salver) *32*,
Guise, Sir W. 54,55	Holford, R.S., MP 58	Linton, Mr 39	34,*41*
Hallow 34	Holland, E., MP 36,57,60	Little Malvern 33	Nunnery Wood Farm............37
Hampshire Downs	Holland-Martin, Lady..........*109*	Llowes Court 22	Oliver, James 20
(sheep) 11,78	Holme Lacy......................16,18	London CC 44	
Hampton Bishop 83	Holmer..................................22	Longhorn............................*110*	Ombersley Agricultural Soc...41
Hanbury 34	Homes, W.H. 74	Longmore, E. 22	Onslow, Mr 36
Handley, James....................22	Hook, Mr 53	Long Service 82,92,94,*109*	Open Field System.................9
Haresfield 55	Hope's Field..........................34	Loughborough 20	Owen Street, Hereford 20
Harrison, McGregor 45	Hope, H.T., MP 54	Lower Wick 37	Oxford Downs (sheep).....11,59,
Hartpury 55,85,*90*	under Dinmore 18	Lugwardine 17	99
Hastings, Glynne ...25,44,83,85,	Hop Pole, Worcester.........33,34	Lyde 22	ram.....................................*67*
91,92,93,96	Horneyhold, Mr....................36	Lygon, General 35,36	Earl of 15,17
Hayes, Mrs *104*	horse sales *104*	Lyttleton, Lord 36,39,115	Oxlease, Gloucester 62,73,84
Hay-on-Wye 25	Hoskyns, Sir H. 17	Machinery Lines 116	Painswick Soc 58
Hayward, C. 60	House of Commons 10,18	Madresfield Farmers' Club 41	Pakington, Sir John 36
D.J. 56,57	Howard's Steam Plough 40	Maiseyhampton *67*	Parker, Harry 56
Haywood, H. 25,43	Hulbert, Mr...........................57	Malvern 39,40,71	Partington, George 24
Healey, Rev 60	Humphries of Pershore.........38	Manchester, Bishop of..........39	Parry, T. Gambier.................58
Henry VIII 16	Hussars, 18th 63	Mann, Mr..............................57	Parsons, Joseph....................55
Henwick Turnpike................38	Improved Black (pigs) 59	Marcle 16	Pauls, Messrs......................*111*
Herbert, E. 35	Inland Revenue....................23	Marden 20	pear varieties.......................33
Jeremiah 34	Italians..................................*80*	Margaret, HRH Princess *109*	'Pedigree' *113*
Hereford Agricultural	Ivingtonbury 22	Marling, T.P. 54	Peel, Sir Robert................21,57
Society ....15 *et seq*	James, T................................*69*	Martin, G.E. 39	Penalt 22
accounts *31*	Jenkinson, Sir George 60	Massey Harris 45	People's Park, Glos 58
farmers..............................*12*	Jersey cows 45	Matthews, T.A. 74	Perdiswell Park 71,72,83
minute book 28,29	Jertin, John 54	Meek, Mr 15	Pershore 33,40
rules 28	Jones, John 17	Meikle, Andrew 10	Peruvian Guano 57
schedule 30	Mr 57	Middle Batten Hall 37	Phillips, Capt Mark.............*109*
Show stand *31*	Joseland, L. 74,96	Midland Bank *108,111*	pig *32*
subscriptions 17	Kedward, Mr.........................17	Military Police *109*	Pitchcroft Racecourse 71,76
Herford and Worcester	Kelmscott 53	Millichap, Mr *91*	Pitts, Mrs E...........................45
Agricultural Society............41	Kemp, John 54	Mills, R................................*103*	Pittville 59
passim	Kent, Duke of........................72	Ministry of Agriculture 72	Pleuro Pneumonia 24
amalgamation 48	Kerr, Col Russell...................71	Munitions...........................46	Ploughing Match 34,35,55,56

ponies....................104,112
pony & trap....................107
Ponsonby, Hon Ashley...........57
Poole, Ald C.W....................67
   Edward....................18
Powell, Rev W.H. 23
Powick 35
Prater, Mr 15
Preece, James....................33
   Mr 17
   of Holmer 22
pre-Show trim 106
Price 99
   James 33
   Joseph 22
   Lady 20
   Mr 18
   Sir Robert....................19,20
   W. Phillips....................58
Princess Royal, HRH 86,108
Purslow, Mr....................15
Queen's Hotel, Cheltenham..63, 69
rabbits 80
Racster, W....................22
Radnor 22,25,45
Ram Inn, Cirencester 53
Ransome's 45
Red Cross....................74,82
Redditch 41
Rennel, Lord 96
Rhode Island Red................111
Rhydd, The....................12,19
Rinderpest 23,37,60
Risling N. 36
Roberts, T. 22
Robertson, Capt 60
Rogers, A. (Judge) 22
Romany display 79
Ross Agricultural Society 24
Royal Agricultural
   College....................57,61
   Society....................57
   Cornwall Show 83
   Counties Show 73
      Agricultural Soc 85
   Glos Hussars Yeomanry 63
   Horse Artillery....................119
   Marines....................59
   Norfolk Show....................83
   Show 26,37,40,57,59,61,71
   South Glos Militia...............38
Royds, Mr 36
RSPCA 74
Ryeland sheep 45
   wool....................17
Rymans 99
St Aldwyn, Lord....................91
Salamander, Fire Float............62
Salway,T.J. 47
   Cup. 47
Scudamore 15

Col 17
Lord 16
(Sec) 73
Seal, R.N. 106
Sebright, Sir Thomas.............37
Segrave, Lord 54
Severn Hall 73,86,117
   River 60
Siam, Crown Prince of...........62
Sichuan 103
Simpson's Farinaceous Food..37
Sivell, Shirley Mrs 104
Sheep Building 103
   dip....................105
Sherborne, Lord 53
Sheriff Hutton 39
Shire Hall, Hereford 17
Shorthorns....10,19,29,38,40,45, 58,59,99
Show Box 99
   Secretaries' Assoc...............85
   scenes....................98
Showyard, 1967 95
Shropshire Down sheep ...11,36, 38,45,59
   and West Midland 24,25
Shrub Hill Station....................37
Slatter, Mr....................57
Smith, Frank 96
   Vassar....................62
   W....................35
   Sir W., JP 37
   William....................74
Smithfield Show....................16
Smithin, Mrs....................40
Smythe, Pat....................92
Smythies, Mr....................19
South Down sheep 59
   Wales Bank 11
Spencers 99
Spetchley....................87
   Road 37
Spithead....................9
Spooner, Mr 35
Spread Eagle, Gloucester ..54,55
Spring Garden Show 86,114
Stallard....................15
   John 38
Stansbatch 22
Star & Garter , Worcester......33, 34,37
Staverton Show 81,83,85
stockmen....................77
Stoke Park, Bristol 61
Stonewall Fields,
   Cheltenham....................63
Stourbridge....................39,40,45
Stourhead 54
Stow Soc 58
Stratford Park 61
Stratton, Mr 57
Stroud Show....................61

Soc 58
Sumner, Hugh 92
Swindon 62
Symonds 15
Talbot 99
   Rev G.C. 57
   George 53
Taylor, H. 25
   Mr 16
Tenbury Agricultural Soc 41
Tench, Mr....................17
Terry, Joe....................91,93
   J.P. 74
Tewkesbury....................60,61,62
Thornton, J. 61
Three Choirs Festival 71
   Counties Agricultural Soc...11
         passim
   Show....................71 et seq
   1922 70
   pre-war....................75
Thinghill....................22
Tiannemen Square 102
Tithe, the....................99
Tombs, Robert....................35
Town Hall, Kidderminster 40
Trinder, D. 56
   E. 56
   Mr 24,60
trophies....................118
trotting race 104
Troy 17
Turner, P. 22
turnouts, 1914 69
Unwin, W.N. 63
Vestey, Lord....................108
Victoria, Queen....................9
Victory HMS....................60
vintage tractors, engines.......117
Wadborough Park 40
Wales, HRH Prince of 108
   (1894)....................62
Walford 22
Walker, John....................22
   J.S. 40
   Rt Hon Peter, MP............106
Walkins, Peter....................20
Walters, J.W. 54
Walwyn, Mr....................16
Wanstead 22
War Agricultural
   Committee 46,47
Ward, Lord 56
Wargent, W. 45
Warner, Mr....................57
Warwick Hall Estate............40
Waterford, Dean of...............58
Waterloo, Battle of 11
Wellesbourne 39
Welles, Edmund 33
Wells, (Judge) 53
Westcar of Bucks, Mr 17

Westhide 17
West of England Bank 11
   Ladies Kennel Club.....86,112
Weston, Stafford 74
Wheeler, E.V. 35
White, J.W. 85
Whitney 20
Wigget, A. 63
Williams, ...99
   Edward....................22
   G.A. 60
   Henry 20
   Jonathan 16
Willison, Lady 108
   Sir John....................108
Winchcombe Society............58
   Edward....................22
Winchester Bushel, the....17,18
Winnington, Sir T.E. 33
Winton, Capt de 58,60
Witcombe, G.F.B. 74
   Jonathan 16
Withers, ...99
   E.V. 63
   Vic & Joy 94
Woodhouse 22
   L.H. 46,84
Wombwell's Menagerie 37
Women's Institute................72
   Land Army....................74
Woodforde, James 113
Worcester Agricultural
   Society ....19 passim
   exhibitors....................35
   Show 32
   Chamber of Agriculture 41 43,44
   City....................20
   Corporation 72
   Race Course 44
Worcestershire County
   Council 35
   Yeomanry 38
World War I 11,63
Worth, Ald R.B. 35
Wotton under Edge 55
Wresthall, W.H. 85
Wrigley, L.C. 71
Yarworth, Mr 17
Yatton 16
Yeld of Broome 25
   T. 20
YMCA 86
Young Farmers Club 82
York 39
   Duke & Duchess 75
Yorke, C. 60
   J.R. 60,61
Zhoa Ziyang 102

Subscribers

Presentation copies

1 The Three Counties Agricultural Society
2 The Rt Hon the Lord Plumb DL, MEP
3 Sir Thomas Dunne KCVO
4 The Rt Hon the Lord Vestey
5 Royal Agricultural Society of England
6 Royal Agricultural College, Cirencester
7 Richard Law
8 Lyn & Jenny Downes

9 John & Helen Lewis
10 Clive & Carolyn Birch
11 Michael James Cupper
12 Michael John Cupper
13 Alan Nott
14 Jennifer Tydeman
15 Frederick Garrington John Lewis
16 Poppy Mary Ann Lewis
17 John David Glenn Lewis
18 David & Sharon Lewis
19 Ed & Nancy de Gorter
20 F.G. Buck
21 D.S. Hay
22 M.R. Allfrey
23 W.R. Allington
23 Mrs B.J. Andrews
25 J.H. Andrews
26 G.H. Baker
27 Mrs Anne Ballard
28-29 R.S. Bannister
30 A.P. Barber
31 J.C. Beckett
32 R.E. Beckett
33 Miss B.E.H. Bell
34 John Bell
35 J.S.B. Bennett
36 R.J.G. Berkeley
37 John Rees Bevan
38 T.S.A. Block
39 J.S. Boulton
40 T.F. Bradstock
41 J.F. Brown
42 Cecil T. Bruton
43 Nigel Bullock
44 W.J. Bullock
45-46 Gilliam Bulmer
47 John S. Burchell
48-49 D.H. Burlingham
50 J.L. Butt
51 J.C. Cairns-Terry
52 Miss Ruth Candy
53 J.D.E. Carson
54 Harry Carter

55 Anthony V. Cassey
56 Jim Chapman
57 Miss A. L. Clark
58 Rev R.J.M. Collins
59 Diana Cooley
60 P.M. Colwill
62 R.I. Creery
61 Basil Savidge
63 Mr & Mrs D.F. Daffurn
64 J.K. Dancer
65 Mr & Mrs A.J. Daniell
66 Robert A. Dare
67 The Rev Canon Ivor L.L. Davies
68 R.C.H. Davies
69 Megan Davies
70-71 J.W.M. Dent
72 Hedley Dodds
73 Patricia Dodsworth
74 Lyn M. Downes
75 A.J. Ellis
76 Mrs F.C. Felton
77 A.R. Fisher
78 R. Forster
79 George Fowler
80 John Frazier
81 Eric G. Freeman
82 Clifford G. Freeman
83 Misses P.J. Dunford & A. Freeman
84 Roger Gershon
85 E.D.C. Gobourn
86 Ms M. Goddard
87 Gerry & Ros Godden
88 Mrs V. Goodbury
89 Mr & Mrs J.P. Goode
90 Robert Green
91 Margaret Greening
92 G.J. Groves
93 Mr & Mrs John Hadley
94 Mrs N.M. Hall
95-96 Stephen R. Harries FRICS
97 Mrs P.M. Harrison
98 J.E. Hart
99 Robin Hatfield

100 Alan J. Hawkes-Reed
101 N.A. Hawkins
102 B. Hervey Bathurst
103 David Hiam
104 Michael Hill
105 T.E.B. Hill
106 Eve Hobill
107 Ian V. Hogg
108 Keith C. Holden
109 Miss C. Holme
110 The Hop Pocket (Crafts)
111 A. Hope
112 P.J. Hoskins
113 C.G. Humphris
114 A.F. Hurran
115 Miss G. Hyatt
116 Mrs T.A.S. Jackson
117 T.V. James-Moore
118 Mrs M.E. Jordan
119 Richard Keen
120 A.G. & J. Laird
121 Mr & Mrs A. Lambert
122 J. Lawrence
123 Richard Law
124 Miss Lettice Lygon
125 Derek Little
126 G.R. Littleton
127 Brian Malpass
128 T.A. Matthews (Solicitors)
129 Jenny Marrison
130 Colin J. Manning
131 Maureen Marke
132 D.W. Martin
133 John Matthews
134 J.O. Mattick
135 D. Mead
136 Davis J. Methven
137 Dr C.A. Miller
138 Hum Morgan
139 Peggy Morgan
140 C.M. Morris
141 Hon Sir Charles Morrison
142 B.K. Moss
143 R.G.R. Mumford
144 M.R. Mumford
145 Mrs M.J. Osborne
146 Mr & Mrs Robin Otter

131

147 M.H. Oughton	177 D.E. Stinton	206 R.L. Williams
148 Michael Overton	178 A. Ruth Stephens	207 R.N. Williams
149 B. Owen	179 G.H. Styles	208 B.J. Wright
150 William H. Palmer	180 Mrs Ruth Sutton	209-210 Rob Yarnold
151 The Smith Family, Peg House Farm	181 Rosemary Tallon	211 John Yorke
152 Michael J. Pengelly	182-184 W.J.B. Taylor	212 R.G. Blake
153 Francis W.H. Perkins	185 Barrington M. Robinson	213 R.A. Bradstock
154 Mrs V.M. Perrigo	186 Mrs E.D. Shepherd	214 Mervyn G. Chance
155 Mrs R.A. Phelps	187 Mrs A. Thyne	215 M.W. Cleaver
156 K.R. Phillips	188 Sally Tippett	216 Anthony M. Davies
157 G.P. Price	189 Jack Threadingham	217 Jack Garside, Lord of Sulby
158 C.J. Pudge	190 Anthony John Tribe	218 A.S. Halls
159 Councillor John S. Pulling	191 David Mark Tustin	219 Mrs Pam Holton
160 Mr & Mrs G.V. Raderecht	192 P.C. Underwood	220 Mrs C. Andrews
161 Clive & Sylvia Richards	193 National Farmers' Union, Telford	221 R.A. Law
162 C.C. Roads	194 P.J. Neal	222 Dr Donald Hunt
163 Miss Adrienne Robson	195 Jean Newbould	223 Mrs H. Houlcky
164 P. Rose	196 Keith & Suzie Usher	224 M.J. Mackinlay MacLeod
165 M.H. Rowse	197 P.L. Walker	225 Sam Meredith
166 Mr & Mrs E.J. Sainsbury	198 V.M. Walker	226 D.J. Sivell
167 W.H.D. Scott	199 S.B. Walker	227 Sarah Stefanou
168 Diana Sharpe	200 Derek Wareham, Chairman Herefordshire County NFU	228 Julia Wright
169 I.L. Law	201 Mr & Mrs W. & J. Weaver	229 W.J. Bullock
170 G.W. Sivell	202 W.E. Wilde & Co Ltd	230 Francis Harcombe
171 F.E. Skinner	203 Mrs D.V. Williams	231 Alan R. Hay
172-174 F.L. Smith	204 Geoffrey Williams	*Remaining names unlisted*
175 T. South	205 R.A. Williams	
176 Richard Spalding		